Infrared and
Raman Spectroscopy

PRACTICAL SPECTROSCOPY

A SERIES

Edited by Edward G. Brame, Jr.

Elastomer Chemicals Department
Experimental Station
E. I. du Pont de Nemours and Co., Inc.
Wilmington, Delaware

Volume I: Infrared and Raman Spectroscopy (in three parts)
edited by Edward G. Brame, Jr. and Jeanette G. Grasselli

ADDITIONAL VOLUMES IN PREPARATION

Infrared and Raman Spectroscopy

(in three parts)

PART C

Edited by

EDWARD G. BRAME, Jr.

Elastomer Chemicals Department
Experimental Station
E. I. du Pont de Nemours and Co., Inc.
Wilmington, Delaware

JEANETTE G. GRASSELLI

The Standard Oil Company
Research and Engineering Department
Cleveland, Ohio

MARCEL DEKKER, INC.   New York and Basel

Library of Congress Cataloging in Publication Data (Revised)

Main entry under title:

Infrared and Raman spectroscopy.

   (Practical spectroscopy ; v. 1)
   Includes bibliographies.
   1. Infra-red spectrometry. 2. Raman
spectroscopy. I. Brame, Edward G., 1927-
II. Grasselli, Jeanette G. III. Series.
[DNLM: 1. Spectrophotometry, Infrared.
2. Spectrum analysis, Raman. QC457 I43]
QD96.I5I53   535'.842   75-32391
ISBN 0-8247-6527-3

27,656

PART C.

MARCEL DEKKER, INC.

270 Madison Avenue, New York, New York 10016

Current printing (last digit):
10 9 8 7 6 5 4 3 2 1

PRINTED IN THE UNITED STATES OF AMERICA

PREFACE

This book is the third one of a new series on spectroscopy. It is Part C of Volume 1 and like the two previous parts is devoted to the fields of infrared and Raman spectroscopy.

The aim and objective of this new series is to cover all the various applications of spectroscopy and to illustrate its usefulness in such important and diverse areas as organic chemistry, inorganic chemistry, polymer chemistry, biological science, environmental science, food, textiles, etc. Through the discussions in the various chapters, the reader will gain a better appreciation and understanding of the different uses to which spectroscopy has been applied. Also, it is hoped that this series will not only provide a practical reference to applications of spectroscopy, but will also encourage a cross fertilization of ideas between applications in the different fields of biology, chemistry, and physics.

Volume 1, Part C contains three chapters. The first one, coauthored by Drs. G. J. Thomas, Jr. and Y. Kyogoku, deals with the applications in biological science. Not only do these authors give a comprehensive coverage of the experimental methods used but they also describe a large number of important applications, giving an extensive account of the use of infrared and Raman spectroscopy in biological systems.

Chapter 12, coauthored by Sandra C. Brown and Dr. A. B. Harvey, covers the applications to polymers. Obviously, if full coverage were given, this topic could easily fill an entire book. However, these authors give the pertinent strengths of the Raman and infrared techniques in the analysis of polymers and describe the prime utility of the techniques in both qualitative and quantitative analyses.

Chapter 13, which is the final chapter in the book and in the first volume, is authored by Dr. Clara D. Craver. It gives a comprehensive coverage of the analysis of surfaces, and illustrates the utility of the Raman and infrared techniques in analyzing various kinds of surfaces. It is a fitting end to the first volume.

Finally, Part C contains cumulative author and subject indexes for all three parts of Volume 1.

This book, the final part of the first volume in the *Practical Spectroscopy* series, illustrates the approximate format and subject coverage for future books in the series. The next volume discusses applications of x-ray spectroscopy, and subsequent volumes will cover applications of mass spectroscopy, nuclear magnetic resonance spectroscopy, ultraviolet-visible spectroscopy, emission spectroscopy, etc., in the various fields of chemistry. These volumes will be published as the material is collected, edited, and prepared for publication on an ongoing basis. There will not be a rigid schedule for publication, and some of the books may be published simultaneously.

One final objective is to bring to the practicing spectroscopist a consistent and practical approach in describing the many and varied uses of spectroscopy. In turn, the chemist, professor, or graduate student will find that this series provides a useful overview on the state of the art in applications of spectroscopic methods.

E. G. Brame, Jr.

CONTRIBUTORS TO PART C

Sandra C. Brown, Arapahoe Chemicals, Inc., Boulder, Colorado

Clara D. Craver, Chemir Laboratories, Glendale, Missouri

Albert B. Harvey, Chemistry Division--Code 6110, Naval Research Laboratory, Washington, D.C.

Yoshimasa Kyogoku, Institute for Protein Research, Osaka University, Suita, Osaka, Japan

George J. Thomas, Jr., Department of Chemistry, Southeastern Massachusetts University, North Dartmouth, Massachusetts

---

*Present address: Food and Drug Administration, Bureau of Drugs, Division of Cardio-Renal Drug Products, Rockville, Maryland.

CONTENTS

CONTENTS OF OTHER PARTS

Part A

An Introduction to Molecular Vibrations, Bryce Crawford, Jr. and Douglas Swanson

Inorganic Materials, Robert L. Carter

Organometallic Compounds: Vibrational Analysis, Walter F. Edgell

Ionic Organometallic Solutions, Walter F. Edgell

Part B

Computer Systems, R. P. Young

Organic Materials, R. A. Nyquist and R. O. Kagel

Environmental Science, Donald S. Lavery

Food Industry, A. Eskamani

Petroleum, P. B. Tooke

Textiles, Gultekin Celikiz

Chapter 11

BIOLOGICAL SCIENCE

George J. Thomas, Jr.

Department of Chemistry
Southeastern Massachusetts University
North Dartmouth, Massachusetts

and

Yoshimasa Kyogoku

Institute for Protein Research
Osaka University
Suita, Osaka, Japan

I. INTRODUCTION

A. General Scope

The purpose of this chapter is (1) to outline in the field of
biological science the kinds of problems to which infrared (IR) and
Raman spectroscopy can provide solutions, (2) to describe the principles and experimental procedures which are employed in the spectroscopic study of biological materials, and (3) to review and discuss a number of recent applications.

The domain of biological science may be viewed broadly to en-
compass a number of specialties which differ greatly in the types
of problems that are encountered and the means employed towards
their solution. However, the techniques of vibrational spectroscopy
have been and probably will continue to be used most widely and ad-
vantageously within those specialties (such as biochemistry, bio-
physics, and molecular biology) that are concerned for the most
part with problems at the molecular level. Therefore, emphasis is
given here to the use of vibrational spectra in the study of molecular structures and interactions of biologically important materials.
The particular classes of biological compounds selected for detailed
discussion reflect largely the special interests of the authors.
Nevertheless, it is hoped that the present treatment will also indicate the potentialities and liabilities of the IR and Raman techniques for applications to other biomaterials, including studies
which may be pursued in such related disciplines as medicine, clinical chemistry, and the like. Thus it is our intent to provide here
a representative rather than a comprehensive survey of biological
applications of IR and Raman spectroscopy.

For the applications that are discussed in Sec. III, the reference citations are not necessarily exhaustive, though in most cases
we have tried to cover the literature up to 1975. More comprehensive
treatments of some of the subjects discussed here are given in recent
reviews and monographs [1-14]. The book by Parker [1], for example,
offers a wide-ranging review of IR applications in biochemistry,

biology, and medicine but little on Raman applications. The reviews by Koenig [4] and Frushour and Koenig [5], on the other hand, are devoted exclusively to Raman applications.

B. Vibrational Spectra of Biological Molecules

1. Nature of the Data of Vibrational Spectra

The kinds of information provided by IR and Raman spectra to aid in the solution of problems in structural chemistry have been discussed in Chap. 1. For applications to biological materials, however, some limitations are imposed. First, for biological samples only condensed phases are encountered in which molecular rotations are transformed into oscillations of the molecule as a whole and can usually be neglected. The IR absorption spectra or Raman scattering spectra which are obtained from molecules in condensed phases thus consist of bands or lines, respectively, which are devoid of rotational fine structure and are referred to simply as vibrational spectra. Second, the large size and low symmetry of biological molecules often precludes simple or meaningful calculation of their normal coordinates and normal modes of vibration. As a consequence, the assignment of the observed vibrational frequencies to motions of specific nuclei or molecular subgroups usually rests upon empirical correlations with spectra of model compounds. Third, biological molecules are frequently investigated in the solution state, either aqueous or nonaqueous, and the effects of the solvent on the bands or lines of the solute must not be ignored. When a solvent is employed, some vibrational information is invariably sacrificed as a result of interfering absorption or scattering by the solvent.

In other respects the kinds of information obtained from vibrational spectra of biological molecules are the same as for smaller molecules, namely, the frequencies of intramolecular or intermolecular vibrations in the range 10 to 4000 cm^{-1} (0.3 to 120 terahertz), the intensities of the spectral bands or lines associated with these frequencies, and the polarization characteristics of the absorbed IR radiation or scattered Raman radiation.

2. Use of Vibrational Spectra in Biological Science

a. Determination of Primary Molecular Structure. In the majority of biological applications information regarding the primary molecular structure is sought from vibrational spectra. By primary structure we mean the actual covalent bonding of atoms in the molecule as distinguished from weaker forces of attraction between parts of the molecule (secondary or tertiary structures). The latter types of structure, which impart a preferred three-dimensional conformation to the molecule, are particularly important for the function or activity of many biological macromolecules and are discussed in the next section.

Although the data of vibrational spectra can in principle provide detailed information about the geometric arrangement of atoms in the molecule and the interatomic forces, the calculations required to realize such information are restricted in practice to molecules of relatively small size and/or high symmetry. Most often the vibrational spectrum of a biomolecule is compared with spectra of model compounds to reveal the presence or absence of functional groups. In favorable cases information may also be obtained regarding the relative arrangements of such groups in the molecule. Examples are the determination of tautomeric structures of nucleic acid bases, the determination of ionization states of amino acids, the determination of the sites of protonation or deprotonation of nucleic acid bases, the detection of disulfide or sulfhydryl groups in proteins, and the identification of geometric isomers of unsaturated fatty acids.

b. Determination of Molecular Conformation. A biological macromolecule of given primary structure may exhibit two or more different conformational structures depending upon the conditions of temperature, solution concentration, ionic strength, pH, etc. Frequently the various conformers exhibit different vibrational spectra as a result of differences in the orientations of molecular subgroups or changes in hydrogen-bonding interactions between donor and acceptor groups. When this occurs the vibrational spectrum may be used to detect the conformational transition and, when sufficient data are available on model compounds, to identify the molecular subgroups which are involved in the stabilizing interactions of one or another conformer.

A widely studied example is the determination of protein conformation from the different vibrational frequencies which are exhibited by proteins in α-helical, β-pleated sheet, or random chain conformations. The various protein conformations are characterized by different interchain and intrachain hydrogen-bonding interactions of the peptide groups. The perturbation of characteristic group frequencies (so-called amide frequencies) of the peptide unit are now sufficiently well understood that the kinds of protein conformations present may be reliably determined in many cases from the vibrational spectrum (Sec. III.A.1).

Even in the absence of precise correlations between spectra and conformational structures, a conformational transition can usually be inferred from changes in the vibrational spectrum. For example, the changes observed in Raman spectra of nucleic acids as a function of temperature reveal a change in conformation of the nucleic acid chain as well as changes in the orientations of the nucleic acid bases relative to one another, even though the detailed molecular geometry cannot be deduced from the spectra (Sec. III.B.3).

The kinds of secondary and tertiary structures which are frequently studied by methods of vibrational spectroscopy are those resulting from (1) hydrogen-bonding interactions of polar groups, including those of the solvent, if applicable, (2) hydrophobic interactions of aromatic ring substituents, and (3) geometrical changes in covalently bonded groups (e.g., disulfide bridges in proteins or phosphodiester linkages in nucleic acids). The phenomenon of isotope exchange may also be exploited as a probe of molecular conformation by methods of vibrational spectroscopy (Sec. I.B.2).

   c. *Study of Molecular Interactions.* Besides the intramolecular interactions discussed above, associative interactions of biological molecules with other molecules or ions may be conveniently studied by methods of vibrational spectroscopy. Hydrogen-bonding interactions between biologically important small molecules have been most widely examined. Other examples are the parallel superposition of aromatic ring compounds ("stacking"), such as occurs for the purine or pyrimidine bases of a nucleic acid, the binding of metal ions or polar molecules to polyelectrolytes, and the formation of coordination complexes. In such applications, molecular interactions are detected by the perturbations which they produce in the vibrational frequencies or intensities that are associated with molecular subgroups. If the perturbed IR bands or Raman lines are reliably assigned (e.g., by group frequency correlation), then the sites of molecular interaction can usually be deduced from the spectra. In favorable cases it is also possible to measure the extent of interaction quantitatively from the spectral intensities and to evaluate thereby the equilibrium association constants. If the temperature dependence of the equilibrium constants can also be determined, then the heat of association (ΔH) can be calculated [Sec. III.B.1.c. (1)].

    d. Study of Isotope Exchange Kinetics. Measurement of the rate of exchange of labile hydrogen atoms by deuterium can provide information on the accessibility (to solvent molecules) of the exchangeable groups and such information can in turn be related to molecular conformation. Exchange rates of hydrogen and deuterium, provided they are not too fast, are conveniently measured by vibrational spectroscopy since the large difference in relative masses of H and D nuclei will produce a relatively large shift in frequency of IR bands or Raman lines associated with their motions. For example, the amide frequency of nondeuterated proteins at 1550 cm^{-1} contains a substantial contribution from the deformation of the peptidyl N-H group. On deuteration, the corresponding N-D motion is shifted to 1450 cm^{-1}. Thus the relative concentrations of deuterated and nondeuterated groups may easily be determined by measuring the spectral intensities at 1450 and 1550 cm^{-1}, respectively. The rate of exchange may therefore be determined by a comparison of the relative intensities as a function of time.

Another example is the exchange of the C-H group at the 8-carbon ring position of purines, which proceeds slowly in D_2O solutions. Study of the exchange rate in nucleic acids, which is conveniently carried out by Raman spectroscopy, can provide information on the accessibility of the purines and therefore on the conformational structure of purine-containing biopolymers (Sec. III.B.4.a).

Use of hydrogen-deuterium exchange, as well as O^{16}-O^{18} exchange, to study the kinetics of various enzyme reactions by methods of IR spectroscopy have also been described [1].

e. *Qualitative and Quantitative Analyses.* Vibrational spectra are important in biological science, as in other disciplines discussed in this volume, for qualitative and quantitative identifications. Some obvious areas of applicability in biological science are: following the course of a biochemical reaction, the assay of pharmaceutical preparations, the detection of adulterants in foods, the clinical analyses of blood and other body fluids, breath analyses, the detection of environmental pollutants, and so forth. Numerous such applications are reviewed by Parker [1], including analytical studies in areas of enzymology, microbiology, medicine, and related fields.

II. EXPERIMENTAL METHODS

A. Infrared Spectroscopy of Biological Molecules

A general survey of experimental IR spectroscopy has been given by Potts [15]. Discussions of specialized techniques for instrumentation and sample handling in biological applications of IR spectroscopy have been given by Parker [1] and Thomas [2]. A brief discussion of these procedures will be given here.

1. Instrumentation

For biological applications, a high-performance double-beam grating spectrophotometer like the Perkin-Elmer Model 283 or Beckman Instruments Model 4250 is preferred. Less versatile, but suitable for most biological applications, are lower-cost spectrophotometers

like the Perkin-Elmer Model 735B and Beckman Model acculab 4. The
capability to expand or compress wave number and absorbance scales
is essential for many biological studies.

A recent development has been the use of Fourier transform (FT)
IR spectrometers [16-18], such as the Digilab Model FTS-14, for the
detection of very weak or very low frequency IR bands not easily
detected by conventional dispersion-grating spectrophotometers.
This method may be useful for investigating biological molecules in
dilute aqueous solutions since strong absorption by the solvent can
be more efficiently compensated.

Accessories for biological work, such as fixed and variable
thickness cells, thermostatable cells, microsampling devices, internal reflectance devices, and polarizers are also available commercially [2].

Table 1 lists the transmission properties of various window
materials that are commonly employed in IR absorption cells.

2. *Sample-handling Procedures*

An important practical advantage of IR spectroscopy is that it
can be applied to most biological materials regardless of such gross
physical and chemical properties of the sample as phase, color,
turbidity, and the like. With usually simple procedures, nearly any
material can be prepared for IR study, a versatility that has been
a factor in the widespread use of IR spectroscopy in the biological
sciences.

Most frequently encountered in biological research are solid
and aqueous samples. The latter are of obvious importance because
the chemistry of living material takes place in an aqueous medium.
However, it is often desirable to remove the relatively large effects
of hydrogen bonding between water and its solutes so that bonding
between solute molecules themselves may be studied. [See, for example, Sec. III.B.1.c.(1)]. To do this, a solvent is needed which
lacks both donor and acceptor sites for hydrogen bond formation.
Therefore, nonpolar and weakly polar solvents which lack such sites
also find use in IR studies of biological materials.

TABLE 1

IR Window Materials[a]

Material	Useful limit (cm^{-1})	Comments
Glass	>3700, <185	Not suitable for narrow-path cells.
Fused silica	>2500	Not suitable for narrow-path cells.
Sapphire (Al_2O_3)	>1800	Excellent thermal and mechanical properties. Fusible to glass.
LiF	>1400	
CaF_2	>1000	
As_2S_3	>900	Red glass. High n. Unstable >200°C.
BaF_2	>800	
Irtran-2 (ZnS)	>700	Yellow. High n.
InSb	>650	High n.
NaCl	>550	Most common cell window.[b]
Si	700–500, >300	Very high n. High m.p. Low-frequency limit depends on purity, thickness, temperature.
Ge	5200–400	Same comments as for Si.
KCl	>400	[b]
AgCl	>400	High n. Soft and malleable. Darkens in light. Reacts with brass, etc.
KBr	>330	[b]
KI	>280	Very hygroscopic, very soft.[b]
CsBr	>250	Hygroscopic, soft.[b]
KRS-5	>230	TlBr–TlI. Red. High n. Toxic. Easily deformed.
CsI	>185	Soft, hygroscopic.[b]
Polyethylene	700–30	Flexible, weak, low-melting.

[a]Reproduced from Ref. 2, p. 306.

[b]Not usable with aqueous solutions. High n means high refractive index (large reflection losses).

a. *Solids*. The oil-mull and alkali halide-pellet techniques, familiar to the organic chemist, are employed frequently for biological samples. The advantages and disadvantages of these methods have been discussed [1,2,15]. Thin, single crystals may also yield satisfactory spectra but in practice it is difficult to produce a crystal of sufficient cross section to fill the sample area (approximately 2 x 0.5 cm) of an IR spectrometer. Microsampling techniques may be required for smaller crystallites (Sec. II.A.3.b).

Many biopolymers can be cast from solution as a thin film onto a plate of IR window material, which may then be mounted in a spectrometer. Usually about 0.5 mg of the sample is dissolved in aqueous solution, spread over an area of about 100 mm^2 of AgCl, BaF$_2$, CaF$_2$, or other suitable window material, and evaporated to dryness so that a film of approximately 10 μm is obtained. Since the molecular conformation of the biopolymer may depend upon the humidity of the surrounding air, the film is ordinarily maintained in a sealed hygrostatic cell in which a salt solution has been placed to achieve the desired humidity [19]. A cell of this type, first described by Sutherland and Tsuboi [20], is shown in Fig. 1. Samples may also be deuterated in such a cell. For example, if a saturated solution of sodium bromate in D$_2$O is placed in the reservoir of the cell, the relative humidity (r.h.) in the cell is kept at 92% and virtually all protons in the OH and NH groups of the sample are exchanged by deuterium. Numerous IR studies have been made of nucleic acids using such techniques [9].

The structural information derivable from the IR spectrum of a biopolymer may be increased if the molecules in the film can first be oriented in a preferred direction and then the spectrum of the film determined with polarized IR radiation. Oriented films (e.g., of nucleic acids and polypeptides) are made by stroking the wet fibers unidirectionally until dry. The oriented film is then mounted so that the molecules have one specific direction (say, parallel) with respect to the plane of polarization of the radiation and its spectrum is recorded. This direction is then changed by 90° and a

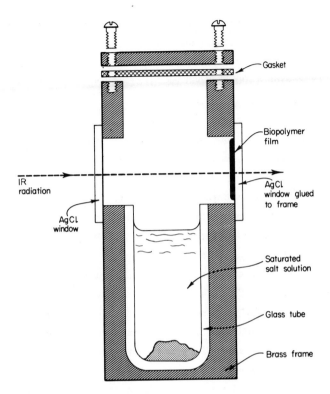

FIG. 1. Diagram of hygrostatic cell for IR absorption spectroscopy of biopolymer films (From Ref. 2.)

second spectrum is obtained. If a particular IR band shows different absorption intensities in the two directions, it is said to be "dichroic."

It should be recognized that a high content of adsorbed water on biopolymer films may lead to the appearance of water bands in the IR spectrum. Conversely, the absence of such bands from the spectrum could indicate that equilibrium with the surrounding humidity has not been achieved. Other precautions which should be exercised in obtaining spectra of oriented films are discussed elsewhere [2,9].

   b. Aqueous Solutions. Because of the opacity of liquid water
to IR radiation, special techniques are required to obtain satisfactory spectra of aqueous solutions [2]. Figure 2 shows spectra in
the mid-IR region (400 to 4000 cm^{-1}) of liquids H_2O and D_2O. It
can be seen that the appreciable absorption of the solvent will obscure much of the vibrational spectrum of the solute. However, with
proper reference beam compensation and with the use of H_2O and D_2O
as companion solvents, most of the mid-IR region can be investigated
[2,9].

   Blout and Lenormant [21] were the first to apply D_2O solutions
to the study of IR spectra of biomolecules, and the technique has
since become widely used for this purpose. Many investigators,
taking advantage of the relative transparency of D_2O in the range
1450 to 1800 cm^{-1} (i.e., the double-bond region, so-called because
of the absorption by C=O, C=C, and C=N stretching vibrations), have
applied the technique in structural studies of proteins and nucleic
acids (Secs. III.A and B). The IR spectroscopy of nucleic acids
and polynucleotides has in fact been developed for the most part
with D_2O as the solvent.

   One consequence of aqueous solution spectroscopy is the loss of
information because of proton exchange with the solvent. For example, NH and OH stretching and deformation modes cannot be observed
in D_2O solutions because of exchange. The resultant ND and OD bands
are in turn obscured by intense D_2O absorption. Thus bands due to
vibrations of hydrogen-bonding donor groups of solute molecules cannot ordinarily be observed in either H_2O or D_2O solutions.

   Cells for aqueous solution spectroscopy may be constructed of
CaF_2 windows (transparent above 1000 cm^{-1}). Other water-resistant
materials are BaF_2, Irtran-2, AgCl, and KRS-5, transparent over
wider ranges than CaF_2 but less convenient to handle (See Table 1).
Ordinarily a cell thickness in the range 25 to 75 μm is most satisfactory when the solute concentration is in the range 100 to 30 mg/ml
(10 to 3% by weight). A minimum volume of about 7 to 20 μl is required to fill the cell. Techniques for more dilute solutions have

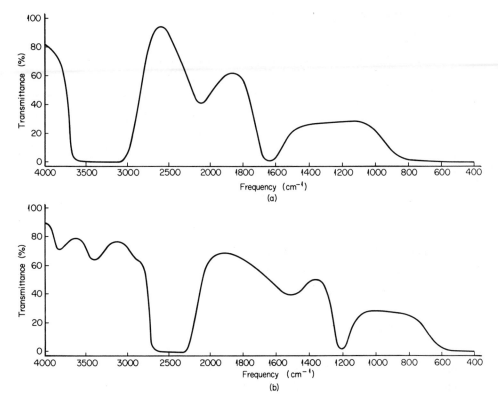

FIG. 2. IR absorption spectra of liquid water in the region 400 to
4000 cm^{-1}. Optical path is 25 μm, AgCl plates. (a) H_2O; (b) D_2O.
(From Ref. 2.)

been described by Miles [22]. The use of acetate or citrate buffers
and the exposure of solutions to atmospheric water vapor should be
avoided since these cause interfering absorption in the double-bond
region; tris, phosphate, and cacodylate buffers are recommended.
In addition, solutes for D_2O solution spectroscopy should be lyphil-
ized first from excess D_2O whenever possible so that proton contamination is minimized.

        Compensation for solvent absorption is best achieved by use in
the reference beam of a variable path cell containing the solvent.
The reference cell thickness is adjusted until a flat base line is
obtained in the spectral range of interest. FT spectroscopy can

also be exploited for efficient compensation of absorption due to the aqueous solvent [17].

   c. *Nonaqueous Solutions.* Details of the use of nonaqueous solvents for solution spectroscopy are given elsewhere [15]. The transmission properties in the mid-IR region of a number of possible solvents of biological materials are summarized in Fig. 3. Of these we shall be concerned here only with chloroform and deuterochloroform for which absorption spectra are shown in Fig. 4. The virtues of chloroform in biochemical studies are its ability to dissolve polar biomolecules (nucleoside derivatives, antibiotics, etc.), its inability to compete for strong hydrogen bonding between solute molecules, and its transparency to IR radiation in the structurally informative OH and NH stretching region (approximately 3400 cm^{-1}). These advantages have been exploited in a series of investigations of the hydrogen-bonding specificity of purine and pyrimidine nucleosides [Sec. III.B.1.c.(1)]. In such studies it is essential to remove from the solvent any polar contaminants which may give absorption bands in the region of interest and which may compete for hydrogen bonding between solute molecules. Passage of $CHCl_3$ or $CDCl_3$ over a column of alumina gel, for example, effectively removes traces of ethanol usually present as a preservative [2].

   In studies of hydrogen bond formation, the OH and NH stretching vibrations that appear near 3400 cm^{-1} can be observed with cells of fused silica windows ("near-IR silica"), available commercially. Other techniques for the preparation, use, and care of IR absorption cells have been described in various places [1,2,15].

   For the study of intermolecular interactions in solutions (nonaqueous and aqueous) of biological molecules, quantitative IR intensity measurements are of critical importance, and these depend upon proper techniques for compensation of solvent absorption. A detailed account of the principles and practice of quantitative IR spectroscopy is given by Potts [15]. Some of the precautions which must be exercised have been discussed previously [2,23].

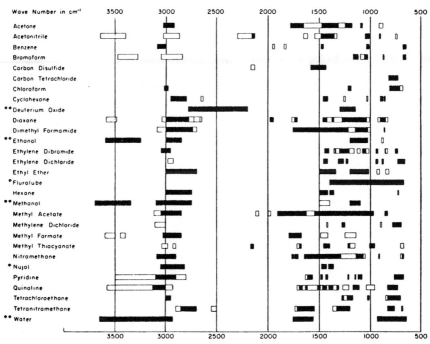

FIG. 3. Transmission properties in the mid-IR region of several common solvents. Optical path is 100 μm, except as stated otherwise. ■ = 0-20% transmission; ▭ = 20-40% transmission; ● optical path undetermined but less than 100 μm; ●● optical path = 15 μm. Chart prepared by Spectroscopy Laboratory, M.I.T., Cambridge, Mass. (From Ref. 2.)

3. *Specialized Techniques*

a. *Measurement of Dichroic Ratios.* In Sec. B.2.a, we have mentioned that the structural information obtained from an IR spectrum of a polymer film may be increased if the molecules in the film are first oriented in a specific direction and the spectrum is recorded with polarized IR radiation. The same applies to other samples for which the vector direction of dipole moment change with respect to the direction of the electric field component of radiation is known (e.g., single crystals). To illustrate this technique we shall discuss its application to oriented films of helical nucleic acids. A similar treatment has been given by Tsuboi [7] and by Hartman et al. [11].

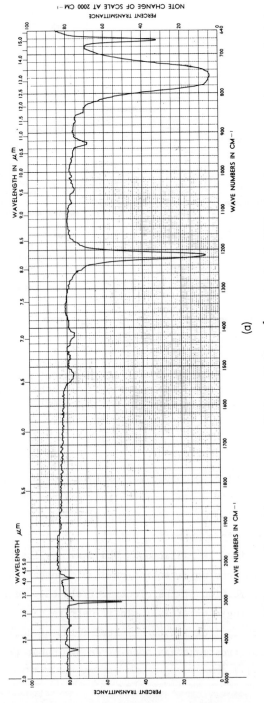

FIG. 4(a). IR absorption spectrum (640 to 5000 cm^{-1} region) of CHCl$_3$ in 10-µm cell. (From Ref. 2.)

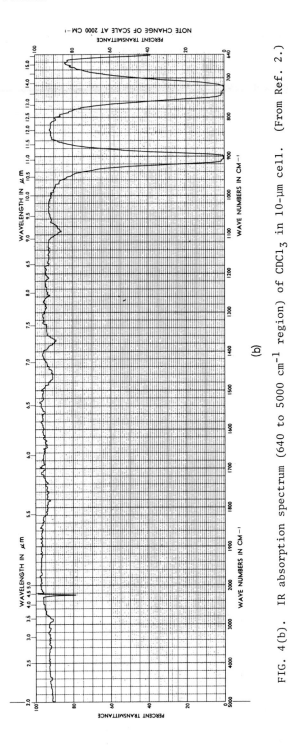

FIG. 4(b). IR absorption spectrum (640 to 5000 cm^{-1} region) of CDCl$_3$ in 10-μm cell. (From Ref. 2.)

A simple and efficient device for polarization of IR radiation
is the polarizer based upon the principle of Brewster's angle [15].
Models are available commercially to fit most IR spectrophotometers.
The grid wire polarizer has also been used effectively for this purpose [7].

The dichroic ratio (R) at a given frequency is the ratio of the
absorbance obtained with the electric vector of the radiation first
parallel ($E_{||}$) and then perpendicular ($E_{\perp}$) to the direction of orien-
tation of the axis of the helical molecules.

$$R = \frac{E_{||}}{E_{\perp}} \tag{1}$$

From the value of R calculated at the frequency of maximum absorption
of a given IR band, the angle θ between the transition moment and
the axis of the helical molecule is given by Falk et al. [24] as

$$\tan^2 \theta = 2 \left[\frac{(1 + f) - R(1 - f)}{R(1 + f) - (1 - f)} \right] \tag{2}$$

which may be rewritten as

$$f = \frac{(R - 1)\cos^2 \theta + (1/2)(R - 1)\sin^2 \theta}{(R + 1)\cos^2 \theta - (1/2)(R + 1)\sin^2 \theta} \tag{3}$$

where f is the fraction of perfectly oriented molecules in the film
and 1 - f is the fraction of randomly oriented molecules. Equations
(2) and (3) also assume that the molecules are helical, so that the
transition moments for a given vibrational mode will sweep out a
cone in space as successive groups are encountered along the helix.
This is a satisfactory approximation for a highly oriented nucleic
acid film.

These equations are applied to double-helical DNA as follows.
At 92% r.h., the planes of the DNA bases are assumed to be perpendicular to the helix axis. The transition moment (M) for the IR band
near 1660 cm^{-1} will by reason of symmetry lie in the planes of the
bases. Therefore, $\theta = 90°$ for the 1660-cm^{-1} band at 92% r.h. A
measurement of R at 1660 cm^{-1} will then give f by Eq. (3). This

value of f may then be used in Eq. (2) to determine θ from values of R obtained at other frequencies (band maxima). This allows the geometric orientations of other functional groups in the molecules to be determined. The value of f may also be used in Eq. (2) to determine θ for the 1660-cm^{-1} band from measurements of R at lower r.h. values. Note that as R approaches unity, however, an ambiguity arises: either θ for the band in question is 55° or the structure has become disordered and randomized.

The results of dichroic measurements on films of DNA and RNA have been summarized in recent reviews [9,11]. The interpretation of these results is, however, open to question in certain cases where adequate precautions were apparently not taken to determine the reversibility of structural changes and the effects of sample heating by the beam of IR radiation. A critical survey has been given by Hartman et al. [11].

 b. Microsampling Techniques. The micro approach generally involves a reduction in the area of illumination (cross section) in the sample beam so that correspondingly smaller samples may be used without reduction of sample thickness. Various devices for achieving this end have been described [1,2,15], including reflecting microscopes which have been used for solid samples of 0.1 µg and microcells for liquid samples of 0.02 µl [1].

 c. Internal Reflection Spectroscopy. Internal reflection spectroscopy (IRS), frequently called attenuated total reflectance (ATR) spectroscopy, is conveniently employed for liquid and solid samples which are difficult to study by conventional transmission spectroscopy. A number of biological applications have been reviewed [1,2]. Parker [1] claims to have found this method particularly suitable for biological specimens, including aqueous solutions as dilute as 5% by weight.

In IRS, the sample is placed in contact with a transparent medium of greater refractive index (internal reflection element) and the reflectance, single or multiple, from the interface between sample and element is measured. A comprehensive study of IRS is given by Harrick [25].

B. Raman Spectroscopy
of Biological Molecules

A number of books have recently been published which deal in
whole or in substantial part with the principles and applications
of Raman spectroscopy [26-31]. At this writing, several other
treatises on applications of the Raman effect, including biological
applications, are in preparation. (See, for example, Refs. 5 and
14.) The reader will find in these works considerable discussion
of the techniques required for the Raman spectroscopy of biological
materials. We shall give here a brief and rather specialized discussion of techniques which contrast the differences between experimental IR and Raman methods, particularly as applied to nucleic
acids and proteins.

1. Instrumentation

The modern Raman spectrophotometer, such as the Spex Ramalog
(Spex Industries, Inc.) or the Cary Model 82 (Varian Associates),
is a double- or triple-monochromator instrument of high dispersion
and low stray-light characteristics. The efficient optical system
of the spectrometer is combined with a powerful laser light source
and a sensitive detector to permit Raman spectra of high quality
(i.e., high signal-to-noise ratio) to be obtained on samples as
small as, or in some cases smaller than, those employed in IR spectroscopy.

Raman spectrometers of lower performance and lower cost, such
as the Cary Model 83 or Spex Ramalab, are of limited usefulness for
biological investigations.

2. Sample-handling Procedures

a. *General Requirements.* The higher cost of Raman over IR
spectrometers is to some extent offset by the lower cost of sampling
devices. For example, sample cells for aqueous solutions are ordinarily made of glass. Powder samples or small crystallites may also
be packed into glass capillary tubes to yield spectra of good quality
and single crystals may be mounted directly in the focus of the
laser beam [2].

In order to obtain a satisfactory Raman spectrum, interaction
of the exciting radiation with the sample by such other mechanisms
as absorption, fluorescence, and Tyndall scattering must be minimized.
Absorption* is eliminated by choice of a wavelength for exciting the
Raman spectrum which is removed from any absorption bands of the sample. The 4880-Å and 5145-Å lines of the argon ion laser are suitable
for most biomolecules. However, biological materials which absorb
in the blue or green may be excited with 6328 Å radiation from the
helium-neon laser. Fluorescence can also be avoided by appropriate
choice of the excitation wavelength. Frequently, the fluorescence
observed in Raman spectra of biological materials is not due to the
biomolecules themselves but to organic contaminants. Such contaminants are often "burned up" after prolonged exposure of the sample
to the laser beam. In other cases the addition of a "fluorescence
quencher," such as potassium iodide, may help to reduce fluorescence.

A related problem in Raman spectroscopy of some biological
materials is the occurrence of a steeply sloping background which is
apparently independent of the chosen wavelength of excitation. This
difficulty has been encountered in studies of aqueous proteins [32]
and nucleic acids [33]. Several explanations have been offered,
including the suggestion that the background is due to electronic
absorption by adsorbed or dissolved oxygen molecules. However,
degassing of solutions does not effectively eliminate the background
in all cases [33]. Elimination of such "background problems" must
await a better understanding of their origins.

Tyndall scattering, usually encountered in solutions, is due to
the presence in the sample cell of suspended particles, such as air
bubbles, dust, colloids, or other undissolved matter, with particle
size comparable to or greater than that of the excitation wavelength.
This type of light scattering detracts from the quality of the Raman
scattering spectrum and is mostly eliminated by high-speed centrifugation or micropore filtration of the sample.

---

*In the case of resonance Raman spectroscopy, an excitation
wavelength is deliberately chosen to be in the region of light absorption by the sample (See Sec. II.B.3.c).

b. Solids. Modern laser Raman spectroscopy permits the routine examination of solids of many types including single crystals,
crystallites or powders, amorphous solids, and polymer films and
fibers. The sample-handling techniques are often dictated by the
specific information sought from the spectra as well as sample
morphology. Few generalizations can be made as regards biological
materials and the reader is best referred to the original literature
for successfully applied techniques. A practical approach is to
review the bibliography of published papers which appears monthly
in *Raman Newsletter* (published by the Optical Society of America,
Washington, D.C., Raman Technical Group).

In the case of biopolymer samples, care must be taken to prevent overheating or photodecomposition of the solid by the laser
beam.

c. Aqueous Solutions. One of the major advantages of Raman
over IR spectroscopy for the study of biological materials is that
aqueous solutions may be investigated with little or no solvent
interference over the range 200 to 2000 cm^{-1}. This is seen by comparison of Fig. 5 and Fig. 2. Thus information may be obtained
more easily on both H_2O and D_2O solutions of biological materials
over most of the vibrational spectrum by use of the Raman technique.

Raman spectra of aqueous solutions of low viscosity are obtained
with relative ease. Thus dilute solutions of biopolymers, or solutions of low-molecular weight materials which are nonaggregated, are
investigated in the same manner as applies to solutions of inorganic
or organic chemical materials (see Chaps. 2 and 6). The concentration of solute required to obtain a satisfactory spectrum will of
course depend upon the intrinsic intensity of Raman lines associated
with the vibrational transitions being observed. For example, vibrations of multiply bonded groups (such as C=C, C=N, and C=O groups in
mononucleotides) and of singly bonded groups of heavy atoms (such as
P-O, C-S, and S-S), give prominent Raman lines when solute concentrations are as low as 0.02 M (approximately 0.3% by weight) and virtually noise-free spectra are obtained when solute concentrations are

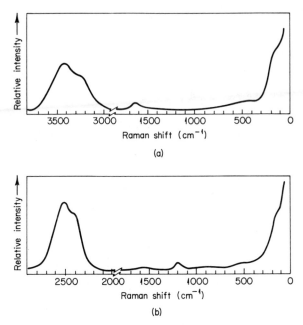

FIG. 5. Raman spectra of liquids (a) H_2O and (b) D_2O. (From Ref. 2.)

0.2 M (3%). On the other hand, for carbohydrates, vibrations of singly bonded groups of light atoms (such as C-C and C-O) give prominent Raman lines only when the solute concentration is about 10% by weight.

Biopolymers, aggregated monomers, and other materials which form solutions of high viscosity may pose formidable problems in sample handling for Raman spectroscopy. One major difficulty is that of loading the cell with a viscous fluid or gel without loss of optical homogeneity. Furthermore, high-molecular weight solutes rarely give clear solutions, so that Tyndall scattering becomes a problem. Considerable progress has been made recently, however, in obtaining Raman spectra of such interesting specimens as guanosine monophosphate gels, highly viscous DNA and RNA solutions, and virus suspensions. (See also Sec. III.B). High-speed centrifugation of the loaded Raman cell is found to be particularly helpful in overcoming optical inhomogeneities in such samples.

A clear solution is not appreciably heated by the laser beam. For example, when 250 mW of 4880 Å radiation is focused on a homogeneous aqueous solution in a glass capillary cell, the solution temperature will usually be in the range 30 to 35°C. However, when suspended particles are present, light scattering may produce heating of the solution. It is not uncommon in such cases to find that severe heating of the sample develops, thus precluding the possibility of obtaining a satisfactory Raman spectrum. Devices for careful control of the sample temperature have been described for both transverse and axial excitation geometries [34,35].

In contrast to modern IR spectrophotometers, which are operated as double-beam instruments, Raman spectrophotometers are usually operated in a single-beam mode. When this is the case, it is not possible to compensate for Raman scattering of solvents, or impurities, by difference spectroscopy. Spectral interference from a solvent is ordinarily compensated by running a spectrum of the pure solvent separately and then subtracting its spectrum from that of the solution, either manually or by computer methods. For quantitative applications, it is usually necessary to make use of an internal standard or to normalize all Raman intensities to a single Raman line of reliably constant intensity [2]. Therefore, quantitative Raman spectroscopy is often more cumbersome than quantitative IR spectroscopy. Recently, a double-beam Raman spectrometer has been utilized to simplify the quantitative measurement of Raman intensities [36].

The intensity of absorption of IR radiation is known to follow the Beer-Lambert law. The intensity of Raman scattering, on the other hand, is not simply related to the optical path. (See, for example, Refs. 37-40.) For a given cell assembly and excitation geometry, however, the intensity of Raman scattering is usually found to be proportional to the volume concentration of scattering molecules. This proportionality factor is often referred to as the "scattering coefficient." A more detailed discussion of the subject of quantitative Raman intensities has been given by several authors [41-44].

3. Specialized Techniques

a. *Polarized Scattering and Depolarization Ratios.* The technique of measuring the degree of depolarization of Raman scattering from biological materials is the same as for other materials [26, 31]. In practice, however, the depolarization ratios are of limited usefulness since most biomolecules are of low symmetry. An exception is the case of depolarization ratios in resonance Raman spectra of heme groups and related biological chromophores. Here the depolarization measurements can reveal whether or not the Raman scattering intensities are governed by a symmetric or nonsymmetric polarizability tensor [13]. Further discussion is given in Sec. III.A.3).

As in the case of polarized IR absorption (Sec. II.A.3.a), polarized Raman scattering can, in principle, provide an increase in the informational content of the spectra. For the special case of an oriented helical biopolymer, the theoretical aspects of polarized Raman scattering have been treated in detail [38-40] and the angular dependence of Raman scattering from randomly oriented molecules has also been considered [37]. In practice, however, it is considerably more difficult to obtain polarized Raman scattering spectra than polarized IR absorption spectra. Experimental obstacles usually encountered are (1) the loss of unidirectional orientation of the molecules in the sample, possibly associated with the effects of laser illumination itself and (2) the inability of conventional collection optics to discriminate the Raman scattering in a given direction (or plane) from light scattered in other directions. Nevertheless, some recent progress has been made in overcoming these difficulties [38], and polarized Raman scattering from at least one helical biopolymer has proved to be structurally more informative [39,40] than the corresponding nonpolarized Raman spectra.

b. *Microsampling Techniques.* Virtually all laser Raman spectroscopy of solid and liquid materials is carried out on microliter or microgram quantities of sample. In the applications cited below (Secs. III.A and B), solutions are routinely investigated in capillary cells requiring no more than a few microliters of sample; and

less than milligram quantities of the solids are easily positioned
in the focus of the laser beam. In the experience of the authors,
the quantities of biological materials that are required for Raman
spectroscopy are usually comparable to and frequently smaller than
those required for IR spectroscopy.

Methods for obtaining satisfactory Raman spectra on nanogram
and nanoliter quantities of sample have also been described [30,45,
46].

c. *Resonance Raman Effect*. In the ordinary Raman effect the
wavelength (or frequency) of the exciting radiation is assumed to
be far removed from the region of electronic absorption by the sample.
When the exciting frequency enters (or approaches) the region of electronic absorption, a *resonance Raman spectrum* is obtained. Detailed
treatments of the resonance Raman effect are given by Behringer
(Chap. 6 of Ref. 26) and Bernstein [14]. A review of applications
to biological materials has also been given recently by Spiro [13].

Theory shows that as resonance is approached, the Raman scattering intensity of certain vibrational modes increases dramatically.
However, the resonance Raman effect cannot always be observed with
ease because the scattered radiation may be greatly attenuated by
absorption and fluorescence processes [26]. Moreover, photodecomposition of the sample may occur. For biological applications, it
is recommended that the sample be checked for molecular weight in-
tegrity and biological activity after the resonance Raman spectrum
is recorded.

The enormous increase of molecular scattering power which can
occur in the resonance Raman effect makes possible the study of
extremely dilute solutions by comparison with the ordinary Raman
effect. For example, an aqueous solution of hemoglobin containing
10^{-4} M (Fe) yields satisfactory spectra with 5682 Å excitation [13].

Finally it should be mentioned that not all of the vibrational
frequencies of a molecule which appear in the ordinary Raman spectrum
are observed in the resonance Raman spectrum. Generally, the only

Raman lines which profit from intensity enhancement are those corresponding to vibrations which determine the vibrational structure of the electronic absorption band nearest to the exciting frequency.

Instrumentation required for resonance Raman spectroscopy is virtually the same as for ordinary Raman spectroscopy. However, the value of the resonance Raman data can be greatly enhanced by the use of a variable wavelength light source, such as a tunable dye laser, to permit excitation within different absorption bands or within different regions of the same absorption band of a biological chromophore.

III. APPLICATIONS OF INFRARED AND RAMAN SPECTROSCOPY

A. Proteins and Related Compounds

A comprehensive review of applications of IR spectroscopy in structural studies of proteins has been given by Parker [1]. Much on this subject has also been reviewed by Susi [3].

The use of the Raman effect in protein research, though pioneered in the 1930s by Edsall and collaborators, is essentially a development of the late 1960s. At this writing, a comprehensive review of Raman applications is in preparation [5]. Numerous Raman spectra of polypeptides and other model compounds are also discussed in Koenig's recent review [4]. We shall attempt to focus our attention here on recent advances in the use of Raman spectroscopy for structural studies of proteins, at present a very rapidly expanding field of research. In this connection it will be useful to review briefly the advances in protein structure determination which have been contributed by vibrational spectroscopy -- mainly through use of IR spectra.

1. Background of Structural Studies by Infrared Spectroscopy

The IR spectra of polypeptides and proteins exhibit a number of characteristic absorption bands (so-called "amide bands"), associated more or less with the peptide (trans-CONH-) grouping.

Variations in the frequencies and intensities of the amide bands
can be used to draw conclusions regarding the structures and conformations of different protein molecules from the IR spectrum.
Nine amide bands are normally distinguished: amide A, amide B, and
amides I to VII, whose frequency ranges are summarized in Table 2.
A discussion of the controversy which over many years centered
around the assignments of these bands is given by Susi [3]. The
currently accepted view is that only the two highest frequency bands
may be assigned to localized bond stretching vibrations (i.e., amides
A and B to N-H stretching), although Fermi resonance with the first
overtone of amide II is also believed to play a part in the intensities of these amide bands [3]. Calculation of the normal modes of
vibration of the -CONH- group in small model compounds, such as N-
methylacetamide [47], shows that the amide I to VII frequencies do
not correspond simply to vibrations described by a single internal
coordinate. Only the major contributions are listed in Table 2,
column 3. Further discussion of the assignments is given elsewhere
[3].

     Spectra of proteins and polypeptides are, of course, more com-
plex than those of smaller model compounds (such as N-methylacetamide)
since the structural repeat unit, or crystallographic unit cell, may
contain more than one peptide grouping from the same or different
protein chains. As a consequence, splitting of the characteristic
amide bands can occur and this splitting may be significant when
intrachain or interchain hydrogen bonding is present. For example,
in polyglycine I, the true unit cell contains four peptide groups,
two from one chain and two from an adjacent chain, and the amide I
band is split into two components at 1630 and 1685 cm^{-1}. The observed splitting (55 cm^{-1}) is greater than can be accounted for by
a simple model in which interchain interactions are neglected [3].

     An approach which takes into account the perturbations to amide
bands that result from both interchain and intrachain interactions
has been described by Miyazawa [47] and further refinements have
been suggested by Krimm and co-workers [48-50]. For the amide I

TABLE 2

Summary of Frequencies and Assignments of Amide Bands

Band	Usual frequency range (cm^{-1})	Approximate description[a]	
Amide A	∿3300	N-H	(3300)
Amide B	∿3100	N-H str (in Fermi resonance with 2 x amide II)	(3100)
Amide I	1597-1680	C=O str, N-H def, C-N str	(1650)
Amide II	1480-1575	C-N str, N-H def	(1560)
Amide III	1229-1301	C-N str, N-H def	(1300)
Amide IV	625-767	O=C-N def	(625)
Amide V	640-800	N-H def	(725)
Amide VI	537-606	C=O def	(600)
Amide VII	200	C-N tor	(200)

[a]Based upon normal coordinate calculations for the frequencies observed in N-methylacetamide (given in parentheses). Amides IV to VII are out-of-plane modes, i.e., out of the plane of the CONH grouping, and others are in-plane modes. Abbreviations: str, stretching; def, deformation; tor, torsion. See also Ref. 3.

and II modes, the significant interactions are (1) intrachain coupling through the α-carbon atoms, (2) intrachain coupling through hydrogen bonding, and (3) interchain coupling through hydrogen bonding. Thus in polyglycine I, the four -CONH- groups in the unit cell are treated as an isolated system of four weakly coupled oscillators. General expressions for the amide frequencies of such coupled oscillators for various protein conformations have been given [51,52]. Following the customary notation [3], the perturbed frequency $\nu(\delta,\delta')$ of an amide mode is

$$\nu(\delta,\delta') = \nu_0 + \sum D_s \cos(s\delta) + \sum D_s' \cos(s'\delta') \tag{4}$$

where ν_0 is the unperturbed frequency of a (hypothetical) isolated oscillator, δ is the phase angle between motions of coupled oscillators in the same polymer chain, δ' is the phase angle corresponding to coupled oscillators in different chains, and the first and second sums are the perturbations from intrachain and interchain interactions, respectively. The interaction coefficients between s^{th} neighbors, D_s and D'_s, are determined by the potential and kinetic energies of interaction and the sums are over all groups in the unit cell. A further refinement of Eq. (4) has also been suggested [50].

Symmetry considerations allow Eq. (4) to be simplified for ordered protein conformations. For example, in the case of the α helix where interchain interactions may be neglected (i.e., $D'_s = 0$), only two values of δ give rise to IR-active vibrations. These are $\delta = 0$, when all oscillators are in phase, and $\delta = 2\pi/3.6$, when the phase angle corresponds to the angle between successive groups relative to the helix axis. Since hydrogen bonding in the α helix is between every third neighbor, it is reasonable to assume that D_1 and D_3 are the most important coefficients, with D_2 playing a less significant role. Equation (4) then gives the two IR-active components for each amide mode as

$$\nu(0) = \nu_0 + D_1 + D_2 + D_3 \quad (||) \tag{5}$$

$$\nu(\chi) = \nu_0 + D_1 \cos \chi, + D_2 \cos 2\chi + D_3 \cos 3\chi \quad (\perp) \tag{6}$$

where $\chi = 2\pi/3.6$. The frequency given by Eq. (5) corresponds to an IR band exhibiting parallel dichroism, i.e., the direction of transition moment change is parallel to the helix axis. Conversely, $\nu(\chi)$ corresponds to a band exhibiting perpendicular dichroism. Both $\nu(0)$ and $\nu(\chi)$ are also Raman-active as is the frequency $\nu(2\chi)$, obtained by replacing χ by 2χ in Eq. (6).

For the β conformation (antiparallel-pleated sheet structure), which applies to polyglycine I, the use of Eq. (4) gives the following frequencies:

$$\nu(0,0) = \nu_0 + D_1 + D'_1 \quad (||) \tag{7}$$

$$\nu(\pi,0) = \nu_0 - D_1 + D'_1 \quad (\perp, \text{in-plane of sheet}) \tag{8}$$

$$\nu(0,\pi) = \nu_0 + D_1 - D'_1 \quad (||) \tag{9}$$

$$\nu(\pi,\pi) = \nu_0 - D_1 - D'_1 \quad (\perp, \text{out-of-plane of sheet}) \tag{10}$$

Equations (7)-(10) apply for the approximation that only nearest neighbor interactions are important. $\nu(0,0)$ is IR-inactive but Raman-active, while $\nu(\pi,0)$, $\nu(0,\pi)$, and $\nu(\pi,\pi)$ are both IR- and Raman-active. Thus three IR frequencies and four Raman frequencies can in principle occur for each amide mode in a protein structure of the β form. In practice, however, some of the branches are too weak and the frequency separations too small to permit all of the expected components to be observed in either IR or Raman spectra (see, for example, Ref. 38).

In order to use the above and other similar equations to determine protein conformations from observed IR (and Raman) frequencies, the constants D_s must be known beforehand. They are customarily obtained empirically from data on polypeptides of known structure. The actual procedure is exemplified by Susi [3] and the results for amide I and amide II modes are summarized in Table 3, adapted from Krimm [53]. These results have been applied successfully to determine chain conformations of a number of fibrous proteins.

In other less successful applications the limitations of the method are revealed. Thus, (1) the amide bands in IR spectra of proteins are frequently too broad and complex to permit conclusions regarding the components $\nu(\delta,\delta')$ that may be present, (2) the band centers may shift significantly with differences in r.h. despite the absence of significant conformational changes in ordered regions of the protein molecules, and (3) the interaction coefficients (D_s and D_s') may exhibit some dependence upon the identity of side chain residues thereby limiting their transferability between different polypeptides of the same chain conformations. Related complications have been considered by Chirgadze and co-workers [54,55].

When the above technique is applied to proteins in aqueous solution, additional difficulties are encountered. For H_2O solutions, solvent interference in the amide I and II regions is severe, and routine application of the IR method is not possible. However, using specialized experimental techniques, Susi et al. [56] have obtained spectra of aqueous myoglobin and β-lactoglobulin and have concluded

TABLE 3

Amide I and Amide II Frequencies Obtained

from IR Spectra of Polypeptides[a]

Conformation	Frequency component	Amide I (cm^{-1})	Amide II (cm^{-1})
Unordered	ν_0	1658	1520
Nylon 66[b]	ν_N	1640	1540
Antiparallel chain pleated sheet	$\nu_{\|\|}(0,0)$	1665(s)[c]	--
	$\nu_{\|\|}(0,\pi)$	1685(w)	1530(s)
	$\nu_\perp(\pi,0)$	1632(s)	1510(w)
	$\nu_\perp(\pi,\pi)$	1650(vw)[d]	1550(w)
Parallel chain pleated sheet	$\nu_{\|\|}(0,0)$	1648(w)	1530(s)
	$\nu_\perp(\pi,0)$	1632(s)	1550(w)
Parallel chain polar sheet	$\nu(0,0)$	1648(s,$\perp$)	1550(s,$\|\|$)
	$\nu_\perp(\pi,0)$	1632(vw)	1530(w)
α Helix	$\nu_{\|\|}(0)$	1650(s)	1516(w)
	$\nu_\perp(2\pi/3.6)$	1646(w)	1546(s)
Polyglycine II (triple helix)	$\nu_{\|\|}(0)$	1624(vw)	1558(s)
	$\nu_\perp(2\pi/3)$	1648(s)	1531(w)

[a]Abbreviations: w, weak; s, strong; v, very.

[b]A model involving only D'_s interactions.

[c]IR-inactive but deduced from Raman spectra (see text).

[d]Revised from Ref. 53 on the basis of more recent data.

that the amide I and amide II frequencies of α-helical and β-type
proteins are not markedly shifted by the aqueous environment. This
view is also supported by Raman data [5]. For disordered or random
chain structures, on the other hand, the amide I mode is shifted to
∿1656 cm^{-1} and the amide II mode to ∿1570 cm^{-1}, reflecting the hydrogen bonding with water molecules to which the IR frequencies are
sensitive.

Solvent interference is not wholly eliminated by study of D_2O solutions since traces of HOD, which are always present, cause interfering absorption at ~ 1550 cm^{-1}. Further, the frequencies of the peptide group are sensitive to hydrogen-deuterium exchange. Consequently for the deuterated peptide grouping (-COND-), the characteristic amide frequencies (denoted amide I', amide II', etc.) and their dependence on chain conformation must also be determined. More on this subject is discussed by Susi [3].

Thus far little mention has been made of the frequencies of the peptide group which occur in Raman spectra of polypeptides and proteins. In fact, little use has been made of Raman spectroscopy in determining chain conformations of proteins that have not already been studied by IR spectroscopy. There are several reasons for this, not least of which has been the relative ease of obtaining IR data as compared with Raman data on biopolymers before the advent of the laser. However, other factors are also involved. The amide II mode, for example, is intrinsically very weak in the Raman effect and reliable frequency and intensity measurements of amide II are not feasible. The polarization of vibrational transitions is also much more difficult to obtain from Raman than from IR spectra of solid samples. Finally, the Raman data so far obtained on polypeptides provide little additional information on the peptide group frequencies that cannot be obtained from IR spectra. For example, Raman spectra of α-helical polypeptides (poly-L-alanine, poly-γ-benzyl-L-glutamate, poly-L-lysine·HCl, and poly-L-leucine) reveal an amide I mode at 1651 cm^{-1} (assigned to the $\nu_{||}(0)$ vibration) which shows no apparent fine structure. The same information is obtained from IR spectra alone. On the other hand, Raman spectra of polypeptides in the antiparallel pleated sheet structure (poly-L-valine, poly-L-serine, polyglycine-I, and oligo-L-alanine) give an amide I frequency at 1666 to 1668 cm^{-1}, presumably due to the $\nu_{||}(0,0)$ mode, which is IR-inactive but the $\nu_{\perp}(\pi,\pi)$ mode, expected hear 1650 cm^{-1}, is apparently too weak to be observed in the Raman spectra of these polypeptides.

In other respects, the Raman and IR spectra of polypeptides and proteins together provide more information than can be obtained from IR spectra alone [57], particularly as regards the study of aqueous samples [58]. Further, the Raman data are often complementary to the IR data and are necessary when a vibrational analysis of the polypeptide chain is attempted [48,49]. Some examples of applications shall be given in the subsequent sections.

2. Structural Studies by Raman Spectroscopy

a. *Amino Acids and Derivatives*. The early Raman studies of Edsall and co-workers on amino acids have been reviewed previously [2,4]. Laser Raman spectra of H_2O and D_2O solutions of the 20 major amino acids at various pH and pD values have been obtained by Lord and Yu (cited in Ref. 59 and available from those authors on request). Raman spectra of these amino acids in the solid state have been reported by Simons et al. [60], who also compare the characteristic Raman lines of the side chain residues with those observed in spectra of oligo- and polypeptides.

Several conclusions may be drawn from these and related spectra: (1) in the Raman effect, it is possible with appropriate use of an internal intensity standard to relate the Raman intensities due to the amino acid side chains quantitatively to their values in proteins, and thus to employ intensity changes as well as frequency shifts for conformational inferences; (2) the high Raman intensities of certain ring-stretching vibrations of aromatic residues (phenylalanine, tryptophan, tyrosine, and histidine) suggest that these groups may be easily detected even in spectra of dilute aqueous proteins; (3) the disulfide group of cystine gives rise to very intense Raman lines near 500 to 540 cm^{-1} and 650 to 730 cm^{-1}, which are due to S-S and C-S bond stretching vibrations, respectively. Their actual frequencies and relative intensities may provide a basis for determining both the number of S-S links in a given protein sample and the C-S-S-C and S-S-C-C dihedral angles. Correlations with other model compounds have also been made [58,59,61-66]; (4) the sulfhydryl group of cysteine gives rise to Raman scattering near 2575 cm^{-1} assigned to the

S-H stretching vibration, thus permitting cystine and cysteine to
be easily distinguished from one another; (5) as the side chains of
amino acids and oligopeptides grow larger and more complicated, the
Raman lines associated with C-C stretching modes of α- and β-carbon
atoms exhibit a corresponding increase in complexity and a systematic
shift to higher frequency. A number of other generalizations can
be made regarding the localized deformation modes of CH, CH_2, and
CH_3 groups [60]; (6) ionized amino and carboxylate groups can be
distinguished from the nonionized groups by their respective characteristic Raman lines. The Raman spectra indicate that the amino
acids exist in zwitterion form in the solid state as well as in
solution; (7) lattice vibrations in crystalline amino acids are
detected by their strong Raman lines in the region 40 to 200 cm^{-1}.

Several of these points are exemplified in the Raman spectra
shown in Figs. 6 and 7.

b. *Oligopeptides*. Raman spectra of glycine oligomers (di-,
tri-, tetra-, and pentaglycines) and of alanine oligomers (di-, tri-,
tetra-, penta-, and hexaalanines) in the solid state have been obtained and compared with spectra of the corresponding monomers and
polymers to reveal a number of interesting correlations [60,67-69].
Raman lines assigned to skeletal vibrations are quite sensitive to
chain length and in principle may be used to differentiate oligomers
from one another. The most interesting spectral differences in the
alanine oligomer series reported by Simons et al. [60] is in the
observed amide I frequencies, which are conformationally sensitive.

Differences between Raman spectra of solid and aqueous alanine
and glycine oligomers have also been discussed [4]. These data
illustrate the advantage of the Raman effect over IR spectroscopy
in the study of aqueous peptides, but the interpretation of the Raman
data in terms of specific conformational structures is open to question. In fact the Raman data obtained by different investigators
[60,67,68] on presumably identical samples are not in satisfactory
agreement with one another. The effect of the degree of polymerization on the Raman spectra of oligo-benzyl-L-glutamates has also been
discussed [38].

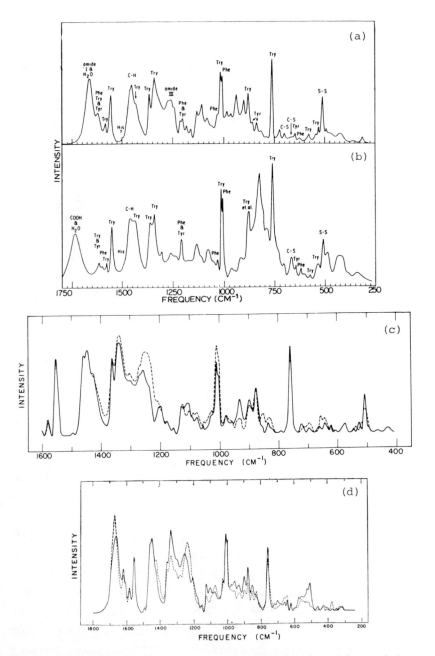

FIG. 6. Comparison of Raman spectra of proteins in native and denatured states. (a) Native lysozyme (H_2O solution, pH 5.2); (b) sum of the constituent amino acids of lysozyme (H_2O solution, pH 1); (c) native (——) and irreversibly denatured (----) lysozyme, normalized to the same scale and corrected for water background; (d) native (——) and S-cyanoethylated (----) lysozyme, normalized to the same scale. Both spectra recorded on solid-state samples. [(a), (b), (c) from Ref. 81, and (d) from Ref. 77.]

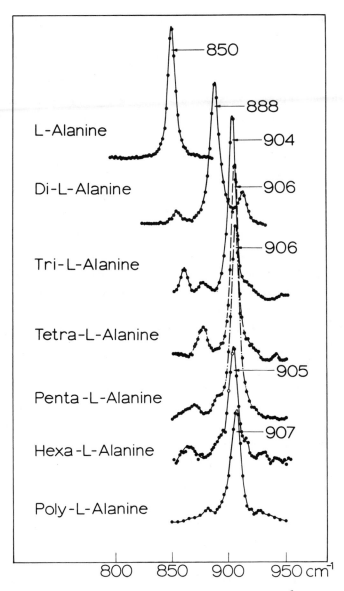

FIG. 7. Raman spectra in the region 800 to 1000 cm^{-1} of the oligo-
alanines. The line at 907 cm^{-1} in poly-L-alanine shifts systematically to 850 cm^{-1} in L-alanine. (From Ref. 60.)

221121

2122

c. Polypeptides. Raman spectra of polypeptides in the solid state and in aqueous and nonaqueous solutions have been reported by numerous investigators. A review of this work in its earlier stages was given by Koenig [4]. Unfortunately, however, some of the more thorough and informative studies of polypeptides [38-40,57,60,70-72] have appeared very recently and therefore are not covered in Koenig's earlier review.

From the larger body of work published to date a better understanding of the Raman spectra of polypeptides has emerged, and indeed a more comprehensive summary of the important amide frequencies and their assignments will appear in print shortly [5]. Rather than reproduce that summary here, we shall give a briefer account which is appropriate to the ensuing discussion on proteins.

Many of the Raman frequencies of polypeptides have been reliably assigned to amide modes of the backbone in one or another conformation, or to vibrations of side chain residues (see, for example, Refs. 40,71,72). However, caution must be exercised when attempting to use these data to draw structural conclusions from Raman spectra obtained on proteins. Some of the limitations of the Raman data in this regard can be seen from the following considerations.

The detection of amide I frequencies of proteins in the Raman effect would appear to be rather straightforward since the lines are relatively intense and little interference from other molecular vibrations is expected in the frequency region of interest (1600 to 1680 cm^{-1}). However, even under such favorable circumstances, components of amide I due to different backbone conformations are difficult to resolve from one another. Usually a single Raman line is observed with no apparent fine structure. Even in the study of oriented films of polypeptides of uniform structure, amide I components of different symmetry cannot be clearly resolved from one another [38,57]. A further complication arises from the fact that the amide I Raman frequencies of polypeptides appear to be sensitive to the method of preparation of the sample or to other experimental factors. A case in point is the α-helical structure of poly-L-alanine, which has been studied independently by several groups [57,60,71,73], three of which report different amide I frequencies. Therefore extreme caution must be exercised in the selection of amide I Raman data from the literature for use in conformational inferences on proteins.

Amide II frequencies of polypeptides (or proteins) are exceedingly difficult to detect in the Raman effect. The low Raman intensity associated with amide II vibrations makes this mode virtually useless for conformational analyses. More often than not [4,5], no Raman scattering whatsoever can be detected in the amide II region (1480 to 1575 cm^{-1}), except that due to vibrations of side chain residues.

Amide III is usually strong in the Raman effect. However in this region (1200 to 1300 cm^{-1}) strong Raman scattering is also anticipated from vibrations of other molecular subgroups. Fortunately, the amide III modes are displaced to significantly lower frequencies ($\sim$900 to 950 cm^{-1}) upon deuterium substitution of the -CONH- groups, by virtue of the substantial contribution of NH bending to this normal vibration, whereas the vibrations of other molecular subgroups are not so greatly affected. Therefore, when it is possible to obtain companion spectra of D_2O solutions of polypeptides, reliable assignment of the amide III modes can be expected. Many of the ambiguities previously surrounding the assignment of amide III modes (cf. Refs. 4,60,73, for example) have now been resolved [5,69], and it appears that the amide III region of the Raman spectrum will be the most useful for conformational studies of proteins.

Finally, the polypeptides are homopolymers containing elements of symmetry, and therefore vibrational selection rules, which are different than those applying to proteins. The different selection rules can in principle invalidate the transferability of vibrational frequencies between the two kinds of polymers. Therefore some caution is required in using spectra-structure correlations valid for polypeptides to make inferences about backbone conformations from the observed Raman spectra of proteins. A similar argument would of course apply to the corresponding IR data.

Despite the difficulties enumerated above, the data from Raman
spectra of polypeptides have been useful in a number of cases for
making deductions about protein conformation when methods other than
Raman spectroscopy could not be used (see Sec. III.A.2.d). Exploita-
tion of Raman spectroscopy for this purpose depends upon the following generalizations, derived from Raman spectra of polypeptide model
compounds [5,71,72]:

1. α-Helical structures give a strong and relatively sharp
Raman line near 1654 ± 5 cm^{-1} (amide I) and a very weak and broad
Raman line near 1285 ± 15 cm^{-1} (amide III). Also associated with
α helices is a Raman line of moderate intensity in the 900 to 950 cm^{-1}
region, which is probably due to a backbone vibration but as yet cannot be more specifically categorized.

2. β structures give a strong and relatively sharp Raman line
near 1670 ± 5 cm^{-1} (amide I) and a strong Raman line near $1235 \pm$
10 cm^{-1} (amide III), occasionally containing resolvable components.

3. Random chain (disordered) structures give a strong Raman
line near 1665 cm^{-1} (amide I) and a Raman line of medium intensity
near 1246 ± 5 cm^{-1} (amide III).

d. *Proteins*. The ambiguities encountered in assigning the
amide frequencies in Raman spectra of polypeptides to particular
conformational structures might have been anticipated from the ex-
periences of IR spectroscopists [3]. The fact that such difficulties
do occur has encouraged the Raman spectroscopist to look elsewhere
in the spectrum for frequencies and intensities that are sensitive
to protein conformation. Fortunately, such structurally informative
regions can be found and, together with the amide frequencies, they
combine to make the Raman spectrum a more versatile source of information on protein structure than the IR spectrum, and a valuable comple-
ment to other physicochemical methods [58].

One region of usefulness, clearly delineated by Lord and co-
workers [58,59,74-77], is that from 500 to 725 cm^{-1}, where strong
Raman scattering occurs from bond-stretching vibrations of C-S-S-C

groups in cystine as well as from C-S groups in methionine and cysteine. The conformational sensitivity of the sulfide group frequencies and intensities in Raman spectra of model compounds and their
applicability to protein spectra have been demonstrated [59,61-65].
A recent application of these results [66] is depicted in Fig. 8
where on the basis of the different Raman frequencies of two proteins
it is possible to conclude that either the trans-gauche-gauche or the
trans-gauche-trans configuration occurs for one of four cystine residues in lysozyme and α-lactalbumin.

     The Raman spectra obtained by Lord and co-workers [59,74,77,81]
on aqueous lysozyme, ribonuclease, and α-chymotrypsin are shown in
Figs. 6, 9, and 10, respectively. In each figure, a composite
spectrum of the constituent amino acids is compared with the protein
spectrum. The S-S and C-S frequencies of cystine groups are assigned
in lysozyme at 509 and 661 cm^{-1} and in ribonuclease at 516 and 659
cm^{-1}. The relative Raman intensities are also greatly different for
the two proteins, suggesting that the conformations of the C-S-S-C
cross links in lysozyme are different from those in ribonuclease and
also from those of the monomeric analogs in solution [74]. The weak
line at 700 cm^{-1} in the spectrum of lysozyme, and the line at 725 cm^{-1}
in the spectrum of ribonuclease are assigned to methionine side
groups [59,74]. Recent studies [63] suggest that the 700-cm^{-1} frequency is characteristic of the trans-gauche configuration and the
725-cm^{-1} frequency of the trans-trans configuration of the methionine
C-C-S-C network.

     It will be noticed that for each of the proteins shown in Figs.
6, 9, and 10, the amide I mode gives a rather strong Raman line near
1660 to 1670 cm^{-1} while amide III gives a complex pattern in the
1200 to 1300 cm^{-1} interval. The conformational assignments suggested
by Lord et al. [59,74-77,81] are summarized in Table 4. These and
related assignments are supported by the observed deuteration shifts
that are detected in the Raman spectra when the same proteins are
dissolved in D_2O solutions and by analogy with spectra of model compounds. As in the case of polypeptides, the amide II modes are of

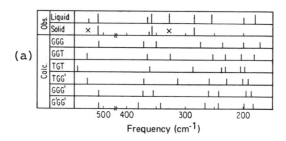

(a)

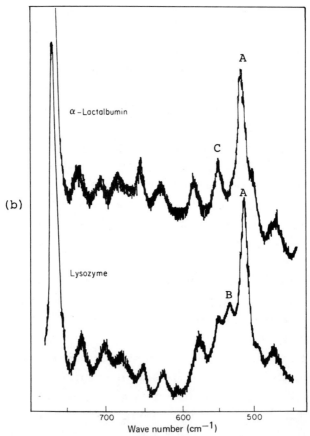

(b)

FIG. 8. Sulfide group frequencies in proteins and model compounds.
(a) Observed and calculated vibrational frequencies of dialkyl disulfide model compounds and their correlation with gauche (G) and
trans (T) configurations of the C-C-S-S-C-C network. (From Ref. 62.)
(b) Application to α-lactalbumin and lysozyme. The strong lines at
507 cm^{-1} (labeled A) are assigned to cystine disulfide bridges in
the GGG configuration. Lines at 528 cm^{-1} (B) and 540 cm^{-1} (C) are
assigned to TGG and TGT configurations, respectively. (From Ref. 66.)

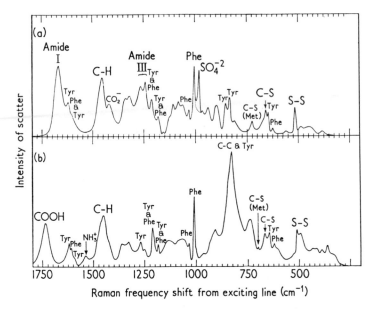

FIG. 9. Comparison of the Raman spectrum (redrawn) of aqueous ribonuclease at pH 5.0 (a) with the sum of the spectra of its constituent amino acids in H_2O at pH 1.0 (b). (From Ref. 74.)

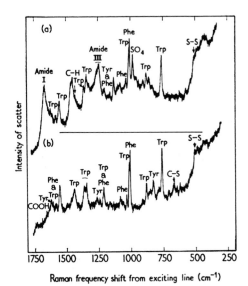

FIG. 10. (a) Raman spectrum of native α-chymotrypsin at 250 mg/ml in H_2O and pH 5.0. (b) Raman spectrum of a mixture of the five constituent amino acids in H_2O at pH 1.0. (From Ref. 74.)

TABLE 4

Amide I and Amide III Frequencies in Raman Spectra
of Aqueous Proteins[a]

	Lysozyme[b]	Ribo-nuclease	α-Chymo-trypsin	β-Lacto-globulin	Bovine serum albumin	Suggested assignments[c]
Amide I	1660	1667	1669	--	--	nonhydrated CONH groups
	1271	--	--	--	1280	α Helix
Amide III	1254	1262	1260	1262	--	Random chain
	1238	1240	1240	1242	1250	β Structure

[a]From Lord et al., Refs. 59, 74-77, 81.

[b]Upon denaturation with 6 M LiBr, the three components of amide III coalesce into a single broad Raman line centered at 1245 cm^{-1}, consistent with the suggested assignments [76,77]. A similar spectral change occurs on S-cyanoethylation [81], Fig. 6.

[c]See also Ref. 71.

insufficient Raman intensity in protein spectra to permit reliable assignment. Further study of the Raman spectrum of lysozyme reveals that the protein chain conformation is essentially the same for crystalline and aqueous samples but substantially different for a lyophilized (amorphous) sample [78] and for denatured samples [76, 77,81]. Figure 6 demonstrates the numerous spectral changes accompanying denaturation of lysozyme by S-cyanoethylation [77].

Detailed Raman spectra of insulin have been reported by Yu and co-workers [79,80,82] for both solid and aqueous samples. These spectra reveal the interesting result that denaturation occurs without cleavage of the disulfide links, since S-S and C-S frequencies are observed in the spectra of presumably denatured insulin [Fig. 11 (a) and (b)]. The frequencies and intensities further suggest that the two interchain disulfide links in denatured insulin assume configurations that are different from those in the native molecule, while the intrachain link, on the other hand, apparently maintains the same geometrical configuration in both native and denatured forms. A comparison of Raman spectra of aqueous and solid insulin·HCl [Fig. 11 (c) and (d)] reveals, further, that the protein structure is not appreciably altered by the aqueous environment. Finally, Yu et al. [80] showed that the insulin moiety in proinsulin gives essentially the same Raman scattering spectrum that is obtained from free insulin, thus suggesting similar chain conformations in the two samples.

Despite the close agreement between Raman spectra of aqueous and solid insulin (Fig. 11), the Raman spectra of aqueous and solid ribonuclease exhibit major differences from one another [83]. These are attributed to changes in the geometry of the disulfide linkages, which also change the local environment of the tyrosine residues but do not significantly alter the backbone conformation [84], a situation quite analogous to that observed for lysozyme [77].

The proteins discussed above have been largely deficient in α-helical chain structure. For comparison, we show in Fig. 12 the Raman spectra of Mojave toxin, a protein which by virtue of its amide

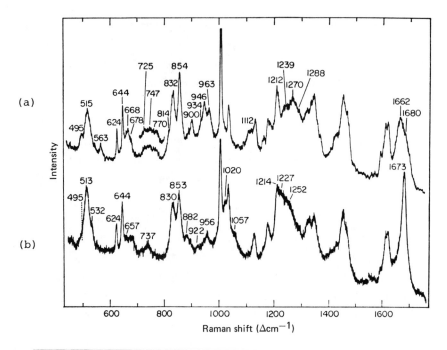

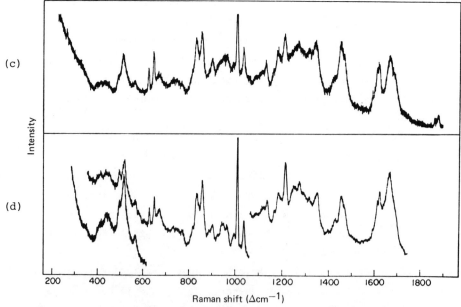

FIG. 11. Raman spectra of insulin derivatives: (a) native zinc insulin, crystalline powder; (b) denatured insulin, heat precipitated from pH 2.4 solution; (c) solid insulin·HCl; (d) aqueous insulin·HCl. (From Ref. 80.)

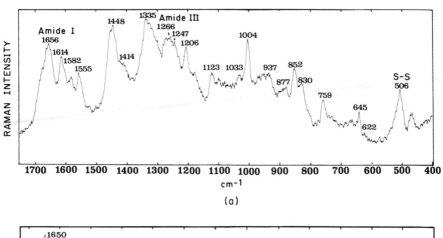

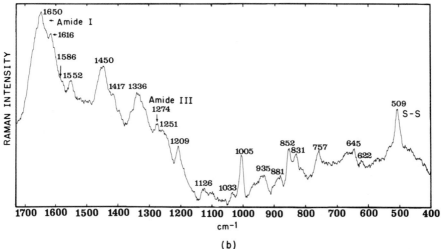

FIG. 12.   Raman spectra of Mojave toxin: (a) solid, (b) H_2O solution.

bands is considered to be largely α-helical.   Unlike the proteins
discussed previously, only weak Raman scattering is obtained in the
amide III region and the observed amide I frequency is close to 1650
cm^{-1} [253].

The information content of Raman spectra of proteins is further
enhanced by the fact that intense Raman lines of phenylalanine,
tryptophan, histidine, and tyrosine are easily detected when these
groups are present (Fig. 6).   These lines, assigned to ring frequencies of the aromatic side chains, respond only minimally to

gross changes in protein conformation [4,77], with one apparent exception. A doublet (825- to 855-cm^{-1} interval), due to tyrosine,
appears to be highly sensitive to the local environment of this residue [83,85,86]. A definitive study of this problem and a highly
useful correlation have now been developed by Siamwiza et al. [259].

     The sulfhydryl group of cysteine generates Raman scattering near
2575 cm^{-1}, which can be detected in Raman spectra of proteins containing this group. Its deuteration shift (to approximately 1875 cm^{-1})
is also easily detected [87]. Therefore, the Raman spectrum provides
a convenient and straightforward means of identifying the presence
of cysteinyl residues (as opposed to cystinyl residues), and in
addition permits conclusions to be drawn about the accessibility of
S-H groups to the solvent by virtue of the observable rate of deuterium exchange. Use of Raman spectra for this purpose should be particularly informative for viral coat proteins where interactions of S-H
groups may be important in stabilizing the virus structure [87].

     A large number of other proteins have also been examined by
means of laser Raman spectroscopy, and an overview is given by
Frushour and Koenig [5]. With the emergence of more reliable assignments and the establishment of sound correlations between observed
spectra and protein structure, it is possible to take advantage of
the Raman effect to determine structural properties of protein subunits in more complex materials, such as nucleoproteins and lipopro-
teins. For example, several viruses have been examined recently to
reveal structural details of their capsid proteins. (See also Sec.
III.B.3.b.)

3. Structural Studies by Resonance Raman Spectroscopy

     The resonance Raman effect is presently being employed in a num-
ber of laboratories to investigate protein structure in dilute aqueous
solutions, i.e., in the range 10^{-3} to 10^{-6} M. Among the materials
studied have been hemoproteins [90-118], their various constituents
[119-128], and nonheme redox proteins such as rubridoxin [90], adrena-
doxin [95], blue-copper proteins [92], and hemerythrin [95]. The
resonance Raman spectra of chlorophyll and its model compound chlorin,

which are closely related structurally to the heme groups, were also obtained [93,99,118]. Studies of other biological chromophores, such as the cyanocobalamins [13], have also been reported. Interactions of proteins with small molecules like azo dyes, which absorb in the visible region of the spectrum, may likewise be studied by the resonance Raman effect [130,131]. Finally, protonation to the Shiff base retinylidene and its linkage with lysine in bacteriorhodopsin have been investigated using this technique [91].

The resonance Raman spectra of hemoproteins have been extensively studied by numerous investigators in order to establish correlations between the spectra and structures of heme groups in proteins as well as to elucidate the mechanism of resonance Raman scattering of the porphyrin nucleus. These results have been recently reviewed [14,105, 256]. Typical of data obtained are the resonance Raman spectra of hemoglobin at different experimental conditions, as shown in Fig. 13(a). In each spectrum the rich pattern of Raman lines in the region 600 to 1700 cm^{-1} arises from vibrational modes of the heme chromophores. The distinct differences in the positions of the Raman lines and in their relative intensities between spectra are attributable to changes in the oxidation and spin states of the central iron in the heme or to ligation or to a combination of these factors. Among the several resonance Raman lines shown [Fig. 13(a)], the intense line between 1535 and 1565 cm^{-1} is most sensitive to the spin state of the Fe ion and the line between 1360 and 1375 cm^{-1} is generally cited as an oxidation state marker [99,101,106,107]. The line of medium intensity near 1490 to 1500 cm^{-1} has also been found as indicative of the structure of the heme group, being related with its spin state [121], while the line at 1615 to 1640 cm^{-1} is sensitive to both the spin and oxidation states [115]. See Table 5.

A recent observation [101] is that reduction of ferric myoglobin to ferrous myoglobin leads to a perceptible lowering of most of the heme vibrational frequencies, which can be explained in terms of "π-back donation" of electrons from the d orbitals of the central iron atom to the π* orbitals of the porphyrin ring. Conversion of iron

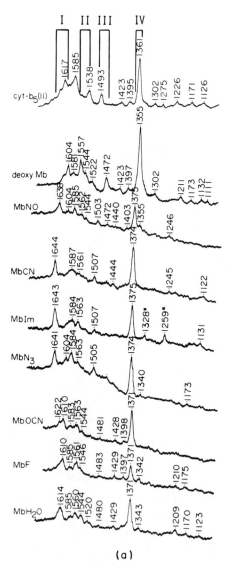

(a)

FIG. 13(a). Resonance Raman spectra of various derivatives of myoglobin and that of ferrous cytochrome $\underline{b_5}$ excited by the 488-nm line. I, II, III, and IV denote the frequency areas of the four key bands. (From Ref. 255.)

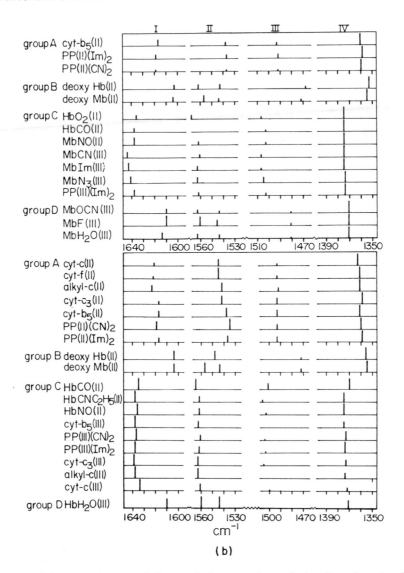

FIG. 13(b). Band intensities and frequencies of the four key bands (I-IV) for the 4880-Å (upper) and 5145-Å excitation lines (lower figure). The strongest line is normalized to unity. (II) and (III) denote Fe^{2+} and Fe^{3+}, respectively. Abbreviations: cyt-c, cytochrome c; cyt-c_3, cytochrome $\underline{c}_3$; cyt-f, cytochrome $\underline{f}$; alkyl-c, dicarboxymethyl methionyl cytochrome $\underline{c}$. (From Ref. 255.)

TABLE 5

Spin States of Fe(II) and Fe(III) in Heme Groups[a]

	Low Spin	High Spin
Fe^{3+}	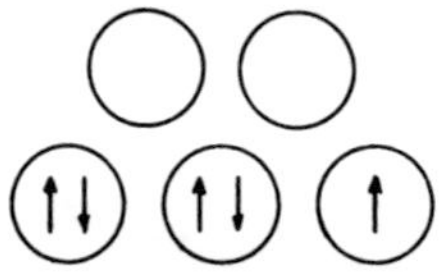 S = 1/2	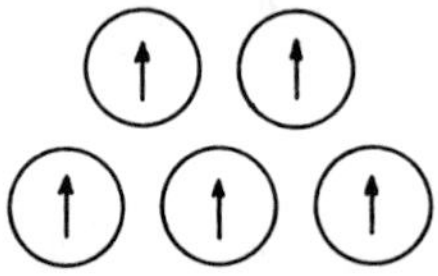 S =5/2
Fe^{2+}	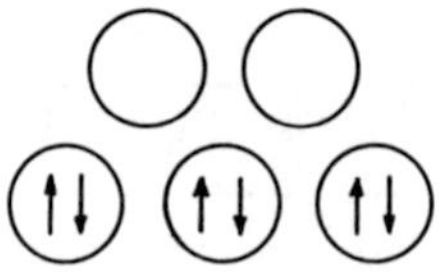 S = 0	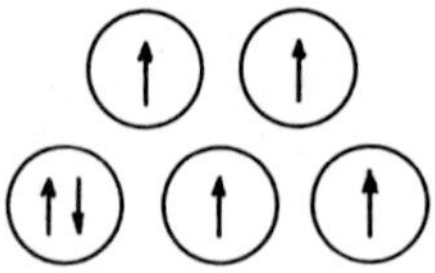 S = 2

[a]Two upper circles in the diagrams indicate d_γ orbitals and three lower circles indicate d_ϵ orbitals.

from high-spin to low-spin states on the other band leads to increases in most of the porphyrin ring frequencies. These increases are attributed to changes in the nature of covalent bonding to the porphyrin ring attendant with a structural deformation which accompanies the change in Fe spin state [101].

Kitagawa et al. [115] classified the spectra of various hemo- proteins containing several different ligands into four groups ac- cording to the positions of their structure-sensitive Raman lines [Fig. 13(b)]. Thus group A contains exclusively ferrous low-spin (S = 0) molecules, while groups B and D cover the ferrous high-spin (S = 2) and ferric high-spin complexes (S = 5/2), respectively. Group C, however, includes two kinds of complexes, namely, the ferric low-spin (S = 1/2) complexes and several ferrous low-spin (S = 0) complexes. It is thus apparent that the kind of spectrum exhibited by *ferric* low-spin complexes can in fact be obtained from complexes which are known from other investigative studies to be *ferrous* low- spin complexes. A possible interpretation of these results is that

all of the complexes (group C members) have in fact the ferric low-
spin bonding between ligand and Fe ion, i.e., O_2, CO, and NO bind to
Fe^{2+} in the same way qualitatively as CN binds to Fe^{2+}. However, to
accept this explanation is to reject out-of-hand the conclusions from
other physicochemical studies (magnetic susceptibility measurements,
^{13}C magnetic resonance spectra, and Mössbauer spectra) which are
apparently reliable. A more reasonable approach would seem to be to
accept the implications of the data of susceptibility, nmr, and
Mössbauer measurements and to postulate that the resonance Raman
effect distinguishes two types of ferrous low-spin hemoproteins,
namely, those of Group A spectra and those of group C spectra. A
satisfactory explanation of how two different spectral patterns can
result from complexes which share the same oxidation state of the Fe
ion and the same spin state of its d electrons is the following.
The difference may lie in the nature of the bonding between the Fe
ion and the so-called "sixth" ligand (axial ligand). Thus for group
A members (cytochromes), the sixth internal ligand is methionine,
histidine, or lysine, each of which may form a σ bond to Fe. On the
other hand, for ferrous low-spin complexes in group C (hemoglobins
and myoglobins), the sixth internal ligand is NO, CO, O_2, or RNC
(alkylisocyanide), each of which may form a π bond to Fe. The extent
of ligand induced "π-back-donation" to the porphyrin rings of the
latter derivatives can be presumed sufficient to account for the near
identity of their spectral lines to those observed for ferric low-
spin complexes. This interpretation is quite reasonable in view of
the electronic structures of the ligands involved [115] and is entirely
consistent with other available data on these complexes.

Several prominent lines in resonance Raman spectra exhibit the
unusual feature of *inverse* or *anomalous polarization,* i.e., the light
scattered at 90° is polarized perpendicular to the direction of polarization of the incident light, when the latter is perpendicular to
the scattering direction. The intensity of the perpendicular component ($I_\perp$) is therefore large while the intensity of the parallel
component ($I_{||}$) approaches zero. Accordingly, the depolarization
ratio, $\rho \equiv I_\perp/I_{||}$, tends to infinity for such Raman lines. This

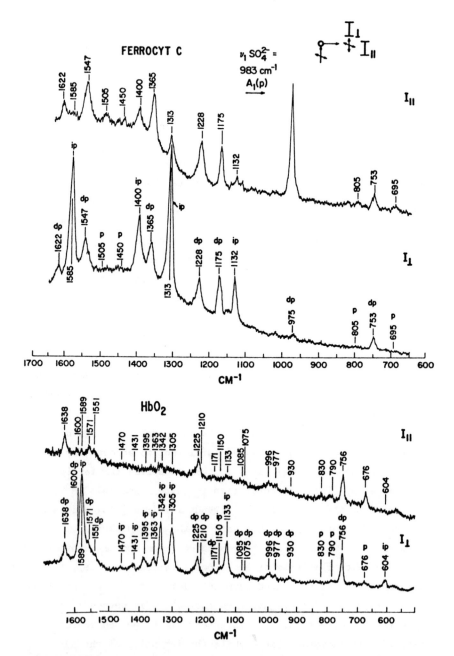

FIG. 14. Polarization properties and frequencies of the Raman lines
in resonance Raman spectra of oxyhemoglobin (bottom pair of curves)
and ferrocytochrome C (upper pair). The scattering geometry is shown

situation is quite different from the normal Raman effect where ρ cannot exceed a value of 3/4 for the same scattering geometry (see Chap. 1). The phenomenon of anomalous polarization is depicted in Fig. 14 for oxyhemoglobin and ferrocytochrome C. The usefulness of such spectra for the investigation of protein structure has been discussed by Spiro and Strekas [100].

Group theoretical considerations show that the approximate D_{4h} symmetry of the planar heme chromophore should give rise to 81 normal modes of vibration, distributed as follows: in-plane, $7A_{1g} + 6A_{2g} + 7B_{1g} + 7B_{2g} + 14E_u$; out-of-plane, $2A_{1u} + 5A_{2u} + 4B_{1u} + 3B_{2u} + 6E_g$. The resonance Raman spectrum is expected to be dominated by those in-plane modes involving C-C and C-N skeletal stretching and methine C-H bending, of which there are $4A_{1g}$, $4A_{2g}$, $5B_{1g}$, $4B_{2g}$, and $9E_u$ (Table 6). The A_{1g} modes are expected to be polarized (p), the B_{1g} and B_{2g} modes depolarized (dp), the A_{2g} modes inversely or anomalously polarized (ip or ap), and the E_u modes Raman-inactive. The spectral data in Fig. 14 thus correlate well with these predictions, assuming several accidental degeneracies in the B_{1g} and B_{2g} vibrations to account for the small number of depolarized Raman lines. Since the side chains attached to the heme group and the interactions with apoprotein in fact lower the symmetry of the heme group, the depolarization ratios are slightly different from those expected for the case of rigorous D_{4h} symmetry. See Table 7.

Inverse or anomalous polarization is due to the anti- or asymmetric scattering tensor which was predicted by Placzek [257]. It has been confirmed with ferrocytochrome C by measuring Raman backscattering with both linearly and circularly polarized light [110, 111].

[FIG. 14 (continued)] schematically in the diagram at the top. Both the direction and polarization vectors of the incident laser radiation are perpendicular to the direction of scattering. The scattered radiation is analyzed into components perpendicular (I⊥) and parallel (I∥) to the direction of polarization of the incident radiation. Excitation wavelengths are 5682 Å for oxyhemoglobin and 5145 Å for ferrocytochrome C and solute concentrations are approximately 5 x 10^{-4} M. Abbreviations: p, polarized; dp, depolarized; ip, inversely polarized. (From Ref. 100.)

TABLE 6

Symmetries of In-plane Skeletal Stretching and

Methine Proton Bending Vibrations of the Porphyrin Nucleus

Internal coordinate	Number of contributions to each symmetry class				
	A_{1g}	A_{2g}	B_{1g}	B_{2g}	E_u
N - C_a	1	1	1	1	2
C_a - C_m	1	1	1	1	2
C_a - C_b	1	1	1	1	2
C_b - C_b	1	0	1	0	1
δC_m - H	0	1	1	1	2
Totals	4	4	5	4	9

TABLE 7

Depolarization Ratios of Raman Lines of Ferrous Cytochromes[a]

ν[b] (cm^{-1})	Cyt-b_5	Cyt-c_3	Cyt-c	Cyt-f	Alkyl-c
1585	1.4	4.3	12.0	11.0	12.0
1493	0.2	0.3	0.3	0.3	0.2
1361	0.2	0.3	0.4	0.5	0.7
1226	0.4	0.4	0.5	0.7	0.6
1171	0.4	0.5	0.6	0.8	0.7

[a]From Ref. 114.

[b]Frequencies for cyt-b_5.

The resonance Raman lines which are sensitive to the spin and oxidation states are also found in spectra of model compounds of the heme group, such as octaalkylporphyrins and their metal- and ligand-substituted derivatives [119-128]. For some of these, crystal and molecular structures have been determined. Accordingly, correlations have been sought between the observed resonance Raman frequencies and the disposition of the metal atom with respect to the heme plane or the Fe-N internuclear distances [101,123].

Spectral assignments have also been supported by recent studies of isotopically substituted hemes and model compounds [124,127]. These assignments, in conjunction with normal coordinate calculations, are useful in identifying the modes of vibration which are sensitive to the spin and oxidation states [127,128]. Modifications in the Urey-Bladley force constants used by Ogoshi et al. [129] and some additional force constants yield frequencies in good agreement with the observed resonance Raman frequencies of octaethylporphyrin. It thus appears that the spin state-sensitive Raman lines near 1630 cm^{-1} (dp) and 1500 cm^{-1} (p) are due to stretching vibrations of the porphyrin inner ring (C_α-C_m and C_α-N bonds), while the spin state-sensitive line at 1560 cm^{-1} (dp) is due to a C_β-C_β stretching vibration. On the other hand, the oxidation state marker at 1365 cm^{-1} (p) is due mainly to a pyrrole ring breathing mode [Fig. 15(a)].

In the lower-frequency region (below 1000 cm^{-1}), there are some bands which are also sensitive to the oxidation and/or spin states of the heme group. The frequencies 750 and 675 cm^{-1} seem to be especially sensitive to the planarity of the chromophore. These can be assigned to in-plane vibrations of the porphyrin ring, probably involving displacement of the four pyrrole rings relative to one another [126]. Detailed studies of the resonance Raman spectra of several hemoproteins have enabled Kitagawa and co-workers [113,114] to find that the band at 1550 cm^{-1} is also sensitive to the identities of the fifth and sixth ligands. In dicarboxymethyl cytochrome c, this frequency is shifted by changing the solution pH from acidic to alkaline values. From the pH shift profile, the pK value was determined to be 7.9. Since the other Raman lines characteristic of the high-spin state do not appear between pH 3 and 11, the observed frequency shift may be due to a change in the nature of bonding of the sixth ligand, namely, lysine [113].

Similarly, a transition from one low-spin state to another, without substitution of the sixth ligand, was observed for ferric cytochrome b_5 [114]. In comparing the positions of the ligand-sensitive

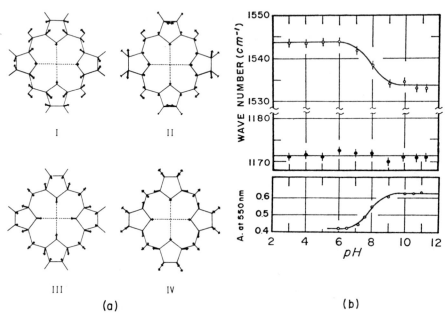

FIG. 15. (a) Modes of the vibrations which give the four key Raman
bands (I-IV), expressed by Lx, the direction and the amplitude of
the displacement of each atom. I, $\sim$1630 cm^{-1}, B_{1g}; II, $\sim$1560 cm^{-1},
B_{1g}; III, $\sim$1500 cm^{-1}, A_{1g}; and IV, $\sim$1365 cm^{-1}, A_{1g}. (From Ref. 128.)
(b) pH-dependent shift of the observed frequencies of the ligand-
sensitive (ϕ) and ligand-insensitive ($\blacklozenge$) Raman lines and pH-dependent
change of absorbance of the α band (peak intensity at 550 nm). (From
Ref. 114.)

lines of cytochrome f and cytochrome c, it can be concluded that the

heretofore unknown ligands of these cytochromes are the same, i.e.,

histidine and methionine [114].

Magnetic susceptibility measurements and nmr spectra of aqueous

myoglobin suggest that there is a "spin equilibrium" in the alkaline

pH region, which allows coexistence of both ferric low- (alkaline

form) and ferric high- (acidic form) spin states. This phenomenon

is clearly confirmed by observing the resonance Raman spectra of

myoglobin at different pH values. Moreover, the resonance Raman

spectra show that the structure of the high-spin form in alkaline

solution is not identical to that of the high-spin form in acidic or

neutral solution [116].

A final demonstration of the utility of resonance Raman spectra in biological science concerns the case of cytochrome oxidase. In this enzyme the A-type heme occurs (i.e., the chromophore contains -CHO in place of the $-CH_3$ group found in B-type hemes). Using resonance Raman spectra, it is possible to show that certain reagents (e.g., $NaHSO_3$ and HCN) react with the chromophoric -CHO group [117].

   4. *Other Applications.*

IR and Raman spectroscopy have been applied in a number of other physical and chemical studies of proteins which will not be reviewed here. These applications include the study of isotope exchange kinetics, mechanisms of enzymatic reactions, and related problems. A comprehensive review has been given by Parker [1].

B. Nucleic Acids and Related Compounds

In the preceding section we have seen that the interpretation of the vibrational spectra of polypeptides and proteins is simplified by an understanding of the normal modes of vibration of the peptide repeat unit -CHR-CO-NH-, where R represents an amino acid-side chain residue. However, the repeat unit in polynucleotides and nucleic acids--a nucleotide--is a more complex structure for which calculation of the normal modes of vibration is a more formidable task. Furthermore, for nucleic acids there are four types of nucleotide residues with significant structural and spectral differences among them. Only very recently have attempts been made to calculate the normal modes of vibration of the nucleotide residues [10,89] and the results of such calculations appear to be of limited usefulness.

On the other hand, considerable effort has been expended to assign to functional groups the IR absorption bands and Raman lines which appear in spectra of the nucleotides themselves and to correlate these frequencies and intensities with those appearing in nucleic acid spectra. The successful use of IR and Raman spectroscopy in structural studies of the nucleic acids is due in large part to the

extensive work on model compounds carried out in a number of laboratories and reviewed previously [2,6-9,11,12]. Several examples of
such investigations shall be discussed in Sec. 1.

   Before proceeding further it will be useful to consider the
structural formulas of the nucleotide constituents of DNA and RNA.
In DNA, the bases adenine, thymine, guanine, and cytosine are con-
nected by a glycosidic linkage to the deoxyribose-phosphate residues.
These nucleotides are depicted in the structures I-IV, where R represents the deoxyribose-phosphate residue. The bases adenine, guanine,
and cytosine also occur in RNA nucleotides but uracil (V) replaces
thymine and ribose replaces deoxyribose. In the polynucleotides and
nucleic acids each monoanionic phosphate group contains two ester
linkages: to the 3' and 5' positions of adjacent sugar residues in
the polymer chain. This is shown schematically for a ribopolymer
(e.g., RNA) in structure VI. In a deoxyribopolymer (e.g., DNA), the
2'-OH group is replaced by hydrogen. The most frequently studied
nucleotides are the 5'-monoesters, which may be obtained from the
polymer (VI) by hydrolysis of the 3'C-O ester bonds.

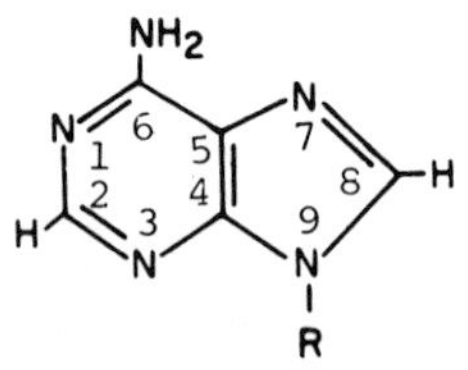

I

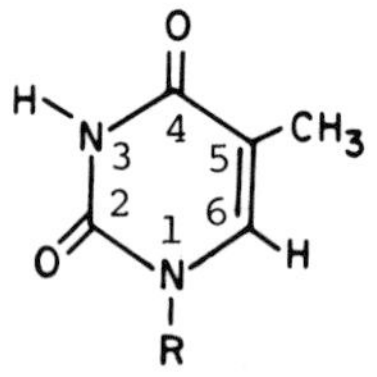

II

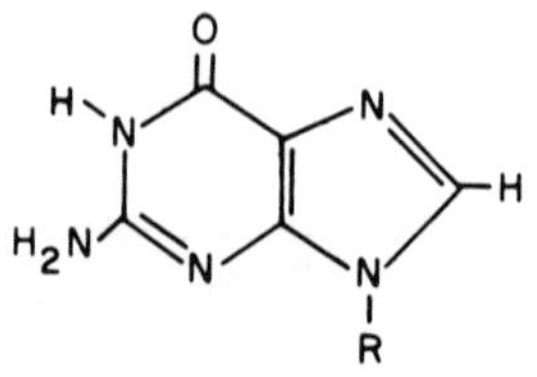

III

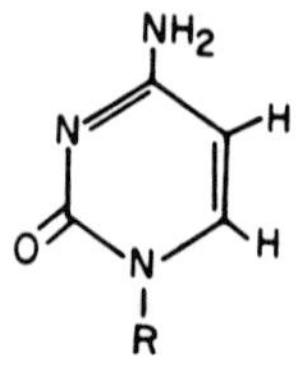

IV

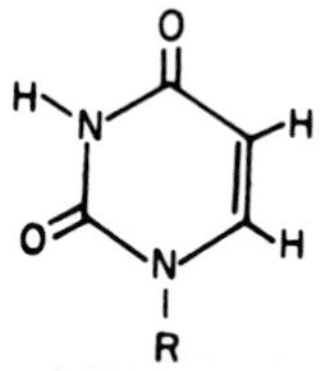

V

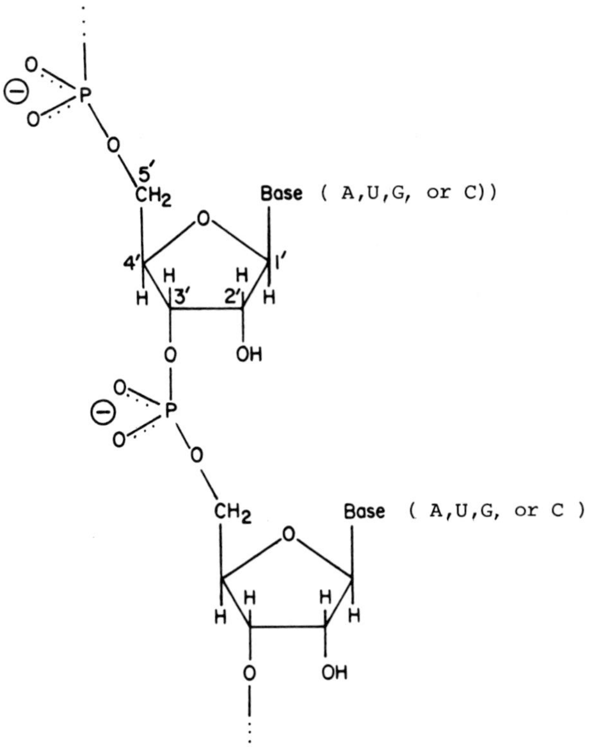

VI

The regions of the vibrational spectrum in which the various nucleic acid constituents give rise to strong IR absorption or intense Raman scattering have been discussed in considerable detail

[6-12]. A brief summary of assignments from spectra of nucleic acids, polynucleotides, and other model compounds is given in Table 8. Interference from H_2O and D_2O, which may be appreciable in spectra of aqueous solutions or of solids at high r.h., are not shown but may be inferred from the IR and Raman spectra of liquids H_2O and D_2O given in Figs. 2 and 5.

1. *Nucleosides, Nucleotides, and Derivatives*

a. Tautomerism. For several years following the proposal of the DNA double helix by Watson and Crick, the question of the prevailing tautomeric structures of the bases, or their nucleosides and nucleotides, remained an issue of considerable debate. Although evidence in favor of the keto and amino tautomers (I-V) was provided by the Watson-Crick scheme of hydrogen bonding and by the results of x-ray diffraction studies of solids, the tautomers of interest are generally those that prevail for the bases *in aqueous solution*. It has been largely by means of vibrational spectroscopy that the actual tautomeric forms of the bases in aqueous solution were determined with reasonable certainty.

Tautomerism is of considerable importance not only with regard to the formation of the Watson-Crick pairs, AT (or AU) and GC, but also for the possible stabilization of unusual base pairs (such as GU) and for the mispairing which may lead to genetic mutations.

As an example of the use of vibrational spectroscopy in elucidating tautomeric structure, we shall consider the hypoxanthine base of inosine, a purine nucleoside which occurs in certain transfer RNA (tRNA) molecules. Inosine may theoretically exist in either the 6-keto (VII) or 6-enol (VIII) form as shown below, where R = ribose.

VII   VIII

TABLE 8

Characteristic Vibrational Frequencies of Nucleic Acids
and Related Compounds[a]

Wave number range (cm^{-1})	Residue	Type of vibration	Relative intensity IR	Raman
3300-3600	Base	ν(NH)	M	W
	sugar	ν(OH)	M	W
3000-3100	Base	ν(CH)	M	W
2800-3000	sugar	ν(CH)	W	W
2300-2500	Base	ν(ND)	W	W
	sugar	ν(OD)	W	W
1600-1750	Base	ν(C=O)	S	M
		δ(NH)	S	W
1550-1650	Base	ν(C=N) ν(C=C)	S	S
1450-1550	Base	Ring stretching	S	S
1300-1460	Base	δ(CH)	W	W
	sugar	δ(CH)	M	M
1250-1450	Base	$\left\{ \begin{array}{c} \text{Ring stretching} \\ \delta(\text{ND}) \end{array} \right.$	M	S
	base		W	S
1225	Phosphate	Antisym ν(O⋯P⋯O)	S	VW
1080-1100	Phosphate	Sym ν(O⋯P⋯O)	M	S
850-1100	Sugar	$\left\{ \begin{array}{c} \nu(\text{CO}) \\ \nu(\text{CC}) \end{array} \right.$	S	W
	sugar		M	W
780-820	Phosphate	$\left\{ \begin{array}{c} \text{Sym } \nu(\text{O—P—O}) \\ \nu(\text{CO}) \\ \nu(\text{CC}) \end{array} \right.$	--	S
	sugar		S	W
	sugar		M	W
650-800	Base	Ring stretching	W	S
300-650	Base and sugar	Skeletal deformations	W	W

[a]Abbreviations: ν, stretching vibration; δ, deformation vibration; sym, symmetrical; S, strong; M, medium; W, weak; VW, very weak; { } indicates vibrations which may be strongly coupled to one another.

Spectroscopic evidence in support of VII was first obtained by Miles [132], who compared IR spectra in the region 1550 to 1750 cm^{-1} of D_2O solutions of inosine, 1-methylinosine, and 6-methoxypurine riboside. The latter two compounds, shown in structures IX and X below, are "locked" by methylations in the 6-keto and 6-enol forms, respectively.

The structure VII for inosine would be expected to produce an
IR spectrum similar to that of IX but different from that of X. Conversely, if inosine has the structure VIII, it should give a spectrum
like that of X but unlike that of IX. The IR data [132] in fact

IX   X

indicate a 6-keto structure (VII) for inosine in aqueous solution.

Despite the IR evidence, the above conclusion was called into
question recently by Wolfenden [133], who argued that keto and enol
forms of inosine exhibited similar basicity (pK_a = 1) and therefore
should be of rather similar stability in aqueous solution. The similarity of u.v. absorption spectra of IX and X was also cited as sup-
porting evidence for this claim. The IR data were considered incon-
clusive on the grounds that the IR band of inosine in question was
broad and was possibly an unresolved doublet, and that the tautomers
VII and VIII could exhibit vibrational frequencies in the double-bond
region which are more closely spaced than those of the methylated
analogs IX and X. The attractiveness of this proposal lay in the
hypothesis that a stable enol structure for inosine could account for
the pairing of I with U in codon-anticodon triplets without invoking
the hypothesis of "Wobble pairing" [133].

It was therefore of interest to reexamine this question by means
of Raman spectroscopy, since the Raman spectra would reveal the vibrational frequencies of the compounds in question not only in the double-
bond region but also in the structurally informative region below
1450 cm^{-1}.

In Fig. 16 the Raman spectrum of inosine is compared with spectra
of the model compounds IX and X. The data obviously confirm the 6-
keto structure. Furthermore, a close examination of the 721-cm^{-1} line
of inosine (Fig. 17) reveals no evidence of a shoulder at 736 cm^{-1},

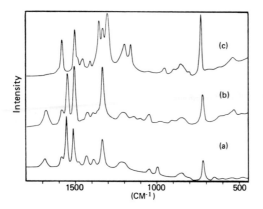

FIG. 16.   Raman spectra of D_2O solutions of (a) 1-methylinosine,
(b) inosine, and (c) 6-methoxypurine riboside.   (From Ref. 134.)

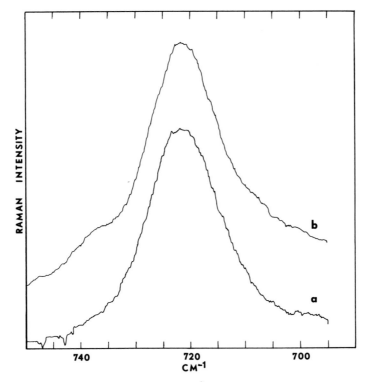

FIG. 17.   Raman lines in the 720-cm^{-1} region of D_2O solutions of
(a) inosine and (b) mixture containing 1 wt % 6-methoxypurine riboside and 99 wt % 1-methylinosine.   Spectra were obtained on an expanded scale with spectral slit width of 5 cm^{-1}.

thus excluding the possibility of significant amounts of the enol form in equilibrium with the keto form. From these results it can be concluded that the equilibrium lies overwhelmingly (>99%) in the keto direction [134].

Data from IR and Raman spectra also support the keto or amino structures for adenine, thymine, uracil, and cytosine residues [135, 136]. However, the arguments favoring the keto-amino structure for guanine are somewhat more tenuous, and the possibility that other guanine ring structures may also coexist in equilibrium with the keto-amino tautomer cannot be excluded for aqueous solutions [137].

b. Ionizations. Protonation* of the bases at low pH (pD) and deprotonation at high pH (pD) produce significant changes in the vibrational spectra. This is illustrated for Raman spectra of adenine and uracil derivatives, shown in Figs. 18 and 19, respectively. The positions of the intense lines assigned to in-plane ring vibrations thus permit the state of ionization of the base residue to be identified from the Raman spectrum [136]. Data such as these also frequently permit the acidic and basic sites of the purine and pyrimidine rings to be identified [6,136,138,139]. For example, the fact that both inosine and 1-methylinosine exhibit the same ring frequencies in their protonated forms identifies either the 3-N or 7-N ring position as the site of attachment [139].

Similar group frequency correlations have led to structural conclusions regarding the sites of protonation and deprotonation in derivatives of other nucleotides [6,136]. A further conclusion from these studies is that ring protonations tend to localize π electrons in the rings at specific single and double bonds, while deprotonation results in delocalization of π electrons [6,136].

At appropriate pH, changes in the state of ionization of the nucleotide phosphate group may also be observed in the Raman spectra. Table 9 lists the observed Raman frequencies, relative intensities, and assignments of phosphate group vibrations for various nucleotides

---

*Protonation will be used to refer to the attachment of either H^+ or D^+. The context will make clear which is intended.

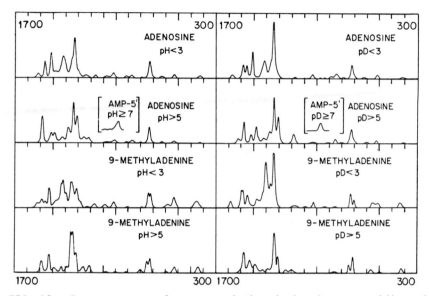

FIG. 18. Raman spectra of aqueous adenine derivatives at acidic and
neutral pH (pD) values. Insets show the additional Raman scattering
from the phosphate group in adenosine-5'-monophosphate. (From Ref.
136.)

in aqueous solution. These frequencies are not easily detected in
IR spectra because of strong solvent interference, but the limited
IR data available confirm these assignments and reveal also a strong
band at 1100 cm^{-1} assignable to the degenerate PO_3^{2-} stretching vibration [6]. Other phosphate group assignments for 2'- and 3'-nucleo-
tides [142] and for cyclic nucleotides [141] have been discussed.

Since the phosphate group gives rise to Raman lines which are
different for different states of ionization and since the lines
occur in a spectral regions which are relatively free of interference
from Raman scattering by either solvent molecules or other functional
groups of the nucleotide, these data may be used to calculate the
acid dissociation constants, as has been demonstrated by Rimai and
co-workers [143,144].

c. Interactions Between Bases. In the secondary structure of
nucleic acids, hydrogen bonds are formed between adenine and thymine
(or uracil) and between guanine and cytosine. The secondary structure

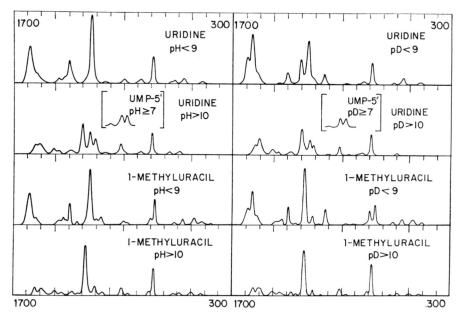

FIG. 19.   Raman spectra of aqueous uracil derivatives at basic and
neutral pH (pD) values.   (From Ref. 136.)

TABLE 9

Raman Frequencies of the Mononucleotide Phosphate Group[a]

Conditions	Group	Frequency (cm^{-1})	Intensity (polarization)	Assignment
pH or pD > 7	$COPO_3^{2-}$	980^{b}	M (P)	PO_3^{2-} Sym str
		1055	W	CO Str
1 < pH or pD < 6	$CO(HO)PO_2^-$	817	M (P)	-O-P-O Sym str
		1050	W	CO Str
		1085	M (P)	$O\cdots P\cdots O^-$ Sym str
pH or pD < 1	$CO(HO)_2PO$	1040	W	CO Str
		1265	S (P)	P=O Str

[a]Observed in Raman spectra of 5'-monoribonucleotides of A, U, G, C,
T, and I (Refs. 136, 139-141).   Abbreviations as in Table 8.

[b]Also detected in IR spectra along with a band at 1100 cm^{-1} assigned
to the degenerate PO_3^{2-} stretching vibration (Refs. 6 and 142).

is also stabilized by van der Waals forces and other short range
forces between purine and pyrimidine rings stacked one above another
along the polynucleotide strands. These interactions, commonly referred to as base-pairing and base-stacking interactions, respectively, have been extensively studied by means of IR and Raman spectroscopy. Since the ability of the DNA (or RNA) bases to interact with
one another is greatly influenced by solvent-solute interactions, it
will be convenient to consider the purine-pyrimidine associations
which occur in nonaqueous (nonpolar or weakly polar) solvents separately from those nucleotide associations which prevail when water
is the solvent.

(1) Studies of Nonaqueous Solutions. In order to examine the
energy of base pairing uncomplicated by solvent effects of comparable
or larger energy, one requires a solvent containing neither donor nor
acceptor groups for hydrogen bonding. The solvent must also be suit-
ably noninterfering in the spectral range of interest in the IR and/
or Raman spectra. Deuterated chloroform ($CDCl_3$) satisfies these requirements rather well. Although equilibrium constants in chloroform
solution and heats of reaction determined from them will differ from
the corresponding gas phase quantities, it is expected that the effect of solvent on the heat of reaction will be small.

The use of IR spectroscopy to detect the hydrogen-bonding interaction in chloroform solution between a purine nucleoside analog,
9-ethyladenine, and a pyrimidine nucleoside analog, 1-cyclohexylura-
cil, was first reported by Hamlin et al. [145]. This study showed
that the purine-pyrimidine association was favored over self-association of either base and that the stoichiometry was that of a 1:1
complex. A quantitative study by Kyogoku et al. [146] later showed
that the equilibrium constants for self-association were smaller by
more than an order of magnitude than that for pairing of A with U.
Moreover, the temperature dependence of the constants gave the heats
of association, $\Delta H°$, -4.0 ± 0.8 for AA, -4.3 ± 0.4 for UU, and -6.2
± 0.6 for AU. Thus the equilibrium constants and heats of association
indicate a clear-cut specificity of the interaction between A and U.
Similar results were obtained for a number of other adenine and uracil
derivatives [147], as shown in Fig. 20.

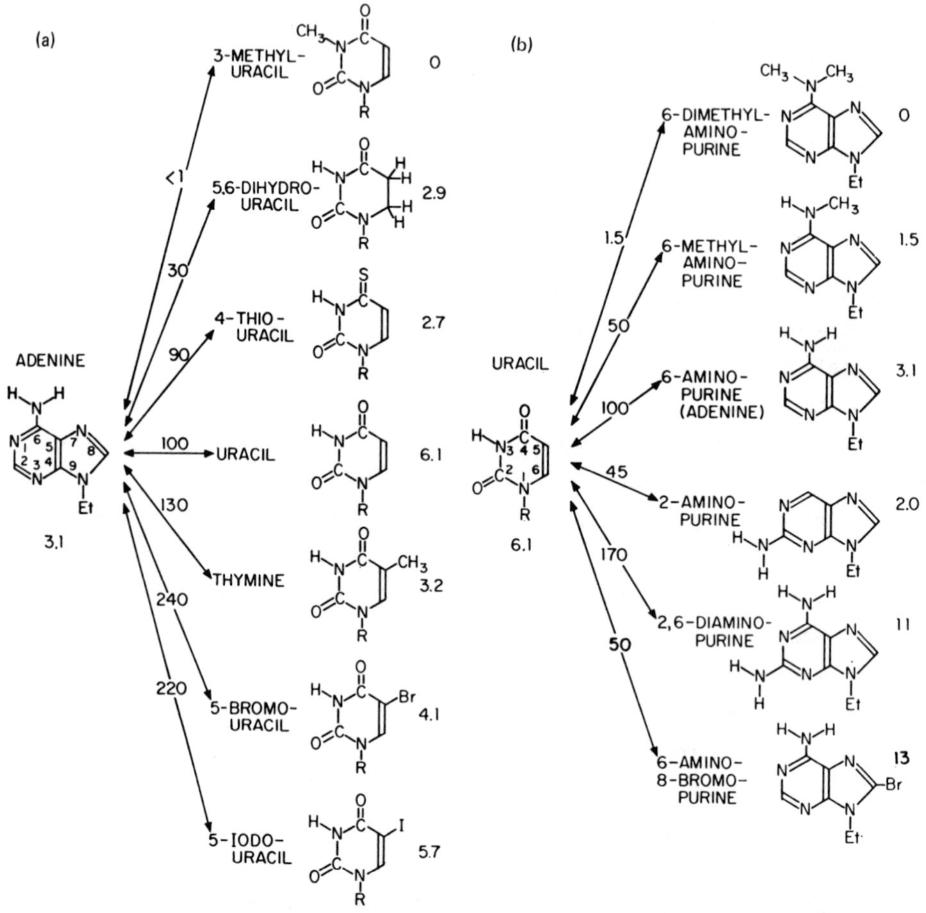

FIG. 20. Association constants (liter/mol) between various derivatives of 1-cyclohexyluracil and 9-ethyladenine in CDCl$_3$ solution at
25°C. Figures near the double-headed arrows are the A-U association
constants and members to the right of the structural formulas are
the A-A and U-U constants. (a) Association of 9-ethyladenine with
uracil derivatives; (b) association of 1-cyclohexyluracil with adenine derivatives. (From Ref. 11.)

It did not prove possible to measure precise values of K for
self-association of nucleoside analogs of guanine or hypoxanthine
nor for the pairing of cytosine derivatives with these purines, for

experimental reasons associated with the large values of K [148]. However, semiquantitative estimates were made and these are summarized in Table 10.

Even where quantitative evaluation of the equilibrium constants is not feasible, it is still possible to demonstrate the *specificity* of base pairing in $CDCl_3$ solution by means of IR spectra of mixtures of the bases. Thus, if no interaction occurs between two bases in solution, the observed IR spectrum of the mixture should be calculable from Beer's law by adding the absorbances of the two bases in separate solutions. On the other hand, if the spectrum of the mixture differs substantially from the calculated spectrum, substantial interaction is demonstrated. For example, a comparison of observed and calculated absorbances reveals that at a total concentration of 0.008 M strong interaction occurs only for the pairs AU, AT, GC, and IC [149]. Thus the postulated base pairing for double-helical DNA

TABLE 10

Approximate Thermodynamic Functions for Base Association
Equilibria in $CDCl_3$ Solution

	$K_{25°}$ (liter/mole)	$-\Delta H°$ (kcal/mole of dimer)	$-\Delta S°$ (e.u.)
C-C	28 ± 3	6.3 ± 0.6	15 ± 1.5
I-I	$4 \cdot 10^2$	$(8)^b$	(15)
G-G	$10^3 - 10^{4a}$	$(8.5 - 10)$	(15)
I-C	$2 \cdot 10^3$	(9)	(15)
G-C	$10^4 - 10^{5a}$	$(10 - 11.5)$	(15)
U-U	6.1 ± 0.6	4.3 ± 0.4	11.0 ± 1
A-A	3.1 ± 0.3	4.0 ± 0.8	11.4 ± 2
A-U	100 ± 20	6.2 ± 0.6	11.8 ± 1.2

[a]Rough estimate.

[b]The figures in parentheses are approximate values based on the assumption of $-\Delta S° = 15$ e.u.

is clearly supported by the specificity of association between the
bases in CDCl$_3$ solution. However, it is not obvious in all cases
where interaction is not observed just why this is so. Thus it is
not clear why A should bind to U and T but not to C and G. As can
be seen from Fig. 20, changes in the ring substituents of A and U
do not destroy specificity for that pair. Strong self-association
of G could suppress its interaction with A and U, but that would not
explain the absence of AC and UC associations. Thus it appears that
in addition to the geometrical requirements for hydrogen-bonding
specificity, there is a specific electronic distribution in the bases
which leads to the specificity of association. Detailed theoretical
studies of the electronic structures of the various bases will per-
haps shed further light on this question.

The association of several barbiturates and related sedatives,
such as glutarimides and hydantoins, with A, U, G, and C derivatives
in CDCl$_3$ was also investigated by IR spectroscopy [150,151]. The
barbiturates display only small self-associations, negligible association with U, G, and C, but a striking specificity for association
with A (for example, K = 1200 liter/mole for the association of
phenobartibal and 9-ethyladenine). These results suggest that the
highly diverse biochemical effects of the barbiturates may be related
to the widespread occurrence of the adenine residue in such substances as ATP and certain coenzymes.

As an extension of this technique, Kyogoku and Yu [152-155]
also demonstrated specific hydrogen-bonding interaction between 9-
ethyladenine and oxidized and reduced forms of riboflavin in CDCl$_3$
solution, between barbital and flavin-adenine dinucleotide and between barbital and nicotinamide-adenine dinucleotide in aqueous
solutions. These studies have been reviewed recently by Kyogoku
et al. [149].

(2) Studies of Aqueous Solutions. In aqueous solution the
strength of hydrogen bonding by the solvent overwhelms that of base
pairing and the fraction of base-paired molecules is too small to
measure by either IR [6] or Raman spectroscopy [156]. However, there

is evidence that purines and pyrimidines do associate in aqueous
solution by stacking of the heterocyclic rings [157], and these
associations may be detected by vibrational spectroscopy.

Tsuboi [9,158] has shown that the intensities of IR absorption
in the region 1300 to 1550 cm^{-1} for the dinucleoside monophosphate
ApG differ significantly from the sum of spectra of the component
mononucleotides. These differences may be attributed to the greater
degree of stacking of the bases in the dinucleoside monophosphate.
Similar differences are observed between the IR spectra of CpU and
its component nucleotides, and these also are attributable to base
stacking.

A more effective way of observing base stacking in aqueous solutions of nucleotides is by Raman spectroscopy. Since the stacking
of the bases in polynucleotides produces diminished intensity of
Raman scattering from certain ring vibrations [Raman hypochromism,
Sec. III.B.2.b.(2)], it is expected that similar hypochromic effects
should be observed for stacking of monomeric base derivatives or for
stacking of bases in lower oligomers, such as the dinucleoside monophosphates mentioned above. For example, small hypochromic effects
have been observed for Ado-5'-P (1.5% by weight in H_2O) at 2°C in
comparison to its spectrum at 80°C [35]. Larger Raman hypochromicities are observed for low-temperature spectra of inosine derivatives
[139], where more substantial base stacking is expected. This is
demonstrated for Ino-5'-P in Fig. 21. Similar results have been
obtained also in Raman studies of aqueous Guo-5'-P and Guo-3'-P
[159,160], although here the nucleotide associations presumably involve hydrogen bonding as well as base-stacking interactions [11].
As expected, still larger hypochromic effects are detected when the
bases of dinucleoside monophosphates are stacked over one another
[161]. Evidence from Raman spectra has also been obtained to indicate that GpC molecules interact with one another in aqueous solution,
presumably with the formation of hydrogen bonds between complementary
bases [161]. In this case, the complex would consist of two base
pairs, apparently providing sufficient energy of stabilization to
overcome the competition from solvent molecules.

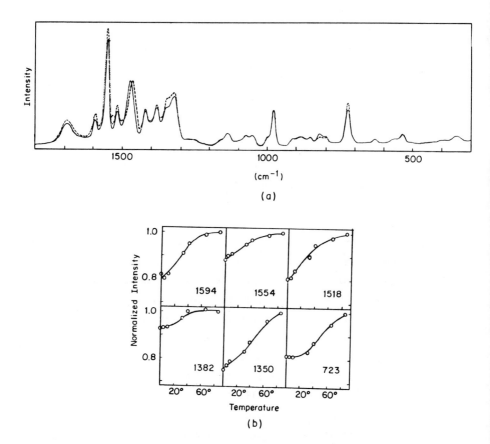

FIG. 21. (a) Raman spectra of inosine-5'-monophosphate (Ino-5'-P)
0.6 M in H_2O solution. Solid curve, 10°C; broken curve, 85°C.
(b) Plots of the Raman intensities vs. temperature for the hypochromic Raman lines at 1594 cm^{-1}, 1554 cm^{-1}, 1518 cm^{-1}, 1382 cm^{-1},
1350 cm^{-1}, and 723 cm^{-1}. (From Refs. 11 and 139.)

        d. Other Interactions. The binding of transition metal ions
to ring nitrogen atoms of the bases has been studied by both IR [162]
and Raman spectroscopy [156,163,164], and the changes produced in the
spectra may be used to measure equilibrium constants of the reactions
involved [11]. The effects of Hg(II), Ag(I), and Au(III) on IR spectra of nucleosides and nucleotides are shown in Fig. 22 and the effects of Hg(II) on the Raman spectra of cytidine are shown in Fig. 23,

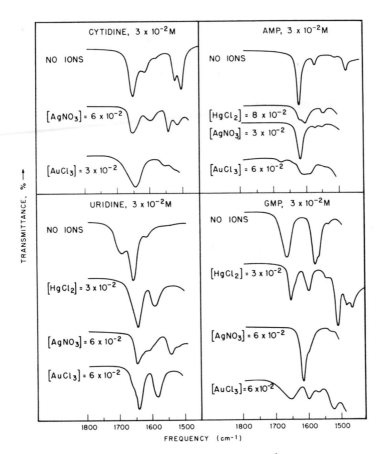

FIG. 22. IR spectra in the 1500- to 1800-cm^{-1} region of D_2O solutions of uridine, cytidine, guanosine-5'-monophosphate (GMP), and adenosine-5'-monophosphate (AMP) complexed with $HgCl_2$, $AgNO_3$, and $AuCl_3$, as indicated. (From Ref. 11.)

to illustrate the type of data obtained. Studies of this sort have benefited greatly from advances in Raman instrumentation which make possible the precise and rapid measurement of spectral differences due to the interactions [36]. A review of other aspects of metal ion nucleoside interactions has also been given recently [165].

The binding of divalent metal ions to nucleotide phosphate groups perturbs the frequencies of vibration of the phosphate group, and these effects are also recognizable in the Raman spectra [139,140, 166,167,182].

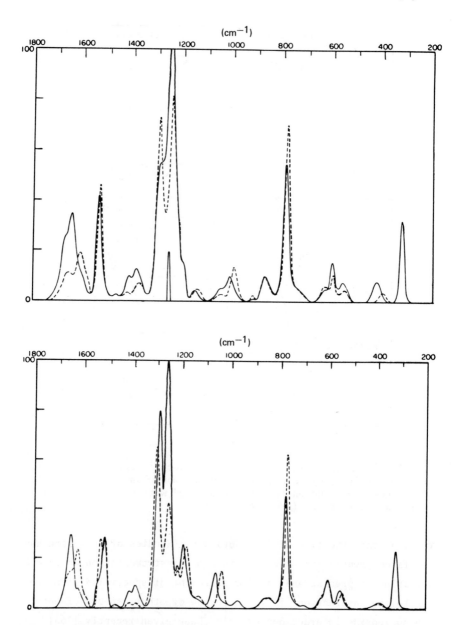

FIG. 23. Raman spectra of aqueous cytidine and cytidine-HgCl2 complexes. Upper pair of curves: solid curve, 0.25 M cytidine in H2O; broken curve, 0.25 M cytidine + 0.25 M HgCl2 in H2O. Lower pair: solid curve, 0.20 M cytidine in D2O; broken curve, 0.20 M cytidine + 0.20 M HgCl2 in D2O. [From Ref. 11].

Kyogoku et al. [168] and Harada and Lord [169] have applied IR
and Raman spectroscopy in the study of crystals of AT and AU dimers
to determine the effects of base pairing on the vibrational frequencies observed for these bases in the solid state. In favorable cases
the hydrogen-bond bending and stretching frequencies can be observed
directly in the solid state spectra.

2. Polynucleotides

a. Features of Vibrational Spectra. The vibrational spectra
of polynucleotides will usually differ from spectra of the corresponding mononucleotides in several respects. First, the frequencies
characteristic of the phosphodiester group $[(CO)_2PO_2^-]$ will replace
those of the phosphomonoester group $(COPO_3^{2-})$ (see also Table 9).
Second, the formation of secondary structure, i.e., base pairing
and/or base stacking, will perturb many of the IR bands and Raman
lines associated with the base residues. Third, in uniformly oriented
fibers, the polarization characteristics of the absorbed or scattered
radiation will differ from that encountered for randomly oriented
monomers in solution. These points have been considered at length
in recent review articles [9,11] and will be exemplified in the foregoing sections.

The frequencies characteristic of the polynucleotide phosphate
group can be observed in IR spectra obtained from thin films of the
polynucleotide deposited on a suitable plate. Thus, poly(rA),
poly(rU), poly(rI), and poly(rC) all give bands at 1230 and 1090 cm^{-1}
assignable respectively to the antisymmetric and symmetric stretching
vibrations of the PO_2^- group [170]. These bands are insensitive to
deuteration and exhibit dichroic ratios which are consistent with
the expected orientations of the phosphate groups in the polynucleotide helices.

In Raman spectra of aqueous polynucleotides the antisymmetric
stretching vibration of the PO_2^- group is too weak to be observed.
However, the corresponding symmetrical mode gives a prominent Raman
line near 1100 cm^{-1} [140,171], consistent with the IR data. In addition, the symmetrical stretching vibration of the phosphodiester

linkages (—O—P—O—) gives a strong Raman line near 810 cm^{-1}, which
is sensitive in position and intensity to the conformation of the
polynucleotide backbone [171,172,186].

Assignment of the IR and Raman frequencies from spectra of
polynucleotides to one or another base residue is rather straightforward, since the bands correspond closely to those observed for
mononucleotides. Further discussion of the assignments and the
polarization properties of the absorbed or scattered radiation are
given in several recent reviews [4,7,9,11,12,140,170].

b. *Conformational Structures and Interactions*. The vibrational
spectrum of a polynucleotide provides a definitive means of investigating its interactions with other polymers, oligomers, monomers, ions,
and the like. Most widely studied has been the formation of complexes
between polynucleotides containing complementary bases, since these
are useful models for secondary structures of nucleic acids. The
pioneering work in this field has been done by Miles and co-workers,
who have employed IR spectroscopy of D_2O solutions of the polynucleotides to detect the formation of complexes, to determine the number
of polynucleotide strands involved, and to investigate thermal stability of the complexes. This work has been reviewed recently by Parker
[1]. In a more recent study, IR spectra and other data have provided
evidence for non-Watson-Crick base pairing between chemically substituted derivatives of poly(rA) and poly(rU) [173].

Use of Raman spectroscopy in the study of polynucleotide interactions has been a more recent development. Although Raman spectra
of polynucleotides were reported in the late 1960s [140], with excitation by mercury arc and helium-neon laser sources, it has been
only with the use of ion laser sources of higher power that this
field has begun to blossom. Surveys of applications are given elsewhere [4,12].

(1) An example: the double-helical complex, poly(rA)·poly(rU).
Advantages of the Raman technique over IR spectroscopy for the study
of aqueous polynucleotides are (1) the accessibility of a wider
region of the vibrational spectrum (300 to 1800 cm^{-1}), which permits

the detection of frequencies not observed in IR spectra and (2) the feasibility of obtaining spectra for both H_2O and D_2O solutions, which permits more reliable assignment of the frequencies to vibrations of specific functional groups. These advantages are revealed in Figs. 24 and 25 which compare IR and Raman spectra obtained on the aqueous polynucleotide complex, poly(rA)·poly(rU) [174-176].

In Fig. 24, the IR spectra of the complex and its products of dissociation at 85°C [an equimolar mixture of poly(rA) and poly(rU)] are seen to be limited to the region 1500 to 1750 cm^{-1}. Nevertheless it is clear that sufficiently large spectral differences exist between the complex and its dissociation products to permit them to be distinguished spectroscopically from one another. Miles and Frazier [174] have exploited this fact in using such IR spectra to follow the thermal denaturation of poly(rA)·poly(rU). Data such as these are also valuable in order to determine, to first approximation, the contributions made by AU base pairs and by unpaired A and U bases to the IR spectra of partly helical RNA molecules [176] (see also

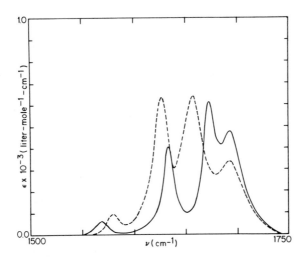

FIG. 24. IR spectra in the region 1500 to 1750 cm^{-1} of D_2O solutions of the double helical complex poly(rA)·poly(rU) at 30°C (solid curve), and the products of dissociation of the complex [equimolar mixture of poly(rA) and poly(rU)] at 85°C (broken curve). (From Ref. 176.)

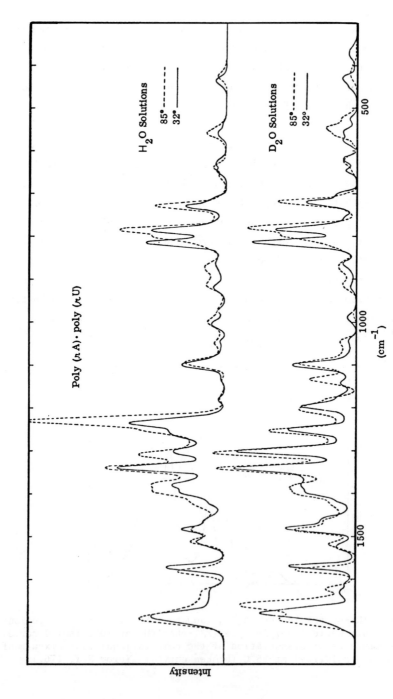

FIG. 25. Raman spectra in the region 300 to 1800 cm^{-1} of H$_2$O and D$_2$O solutions of poly(rA)·poly(rU) (solid curves) and its products of dissociation (broken curves). (From Ref. 12.)

Sec. III.B.3.b.). An important feature of the IR spectra, not common to the Raman spectra, is that light absorption may easily be
measured quantitatively. Thus, the relative absorbances at 1672 and
1657 cm^{-1} in a spectrum of partly dissociated poly(rA)·poly(rU) can
be used, with knowledge of the appropriate molar absorptivities, to
determine quantitatively the fractions of base residues in paired
and unpaired states.

In Fig. 25 the Raman spectra of poly(rA)·poly(rU) and its products of dissociation at 85°C (----) are compared for both H$_2$O and
D$_2$O solutions over the range 300 to 1800 cm^{-1}. For comparison with
Fig. 24 (where solvent interference is compensated by double-beam
spectrophotometry), the background of Raman scattering by the solvent
has been subtracted from the spectra shown in Fig. 25. The data are
also normalized, by use of an internal standard, to permit quantitative comparison of Raman intensities from one spectrum to another
[175]. The Raman spectra thus reveal at a glance the effects of
deuterium exchange (of NH groups in the bases and OH groups in the
ribose moiety) as well as the effects of thermal denaturation of
deuterated and undeuterated forms of poly(rA)·poly(rU). The value
of spectra such as these for investigations of nucleic acids has been
discussed previously [175] and will be reconsidered below (Sec. III.
B.3.b).

(2) Specific Effects of Base Stacking. Of recent interest has
been the question of the effects of base stacking (and therefore of
base sequence) on IR and Raman spectra of polynucleotides. Figures
24 and 25 show that the double-helical complex, poly(rA)·poly(rU),
at 32°C (———) exhibits striking spectral differences from spectra
of mixtures of poly(rA) and poly(rU) at 85°C (----). We may assume
to first approximation that at 85°C the single-stranded polynucleotides are random chains and, therefore, that no appreciable stacking
of the bases occurs. Hence, in the complex all bases are paired (AU)
with each A residue stacked between adjacent A residues and each U
residue stacked between adjacent U residues, while upon dissociation
all bases are unpaired and essentially unstacked. In principle the

perturbations to the spectra of poly(rA) and poly(rU) which result separately from such pairing and stacking interactions could be unraveled by examining suitable model systems in which the bases are (1) paired but unstacked and (2) unpaired but stacked. Unfortunately, for aqueous solutions the first type of model system cannot be found, while the second type is known only for poly(rA). It is therefore not possible to identify unambiguously the spectral changes in Figs. 24 and 25 as being due to either base pairing or base stacking or to both. An approach which can be followed, however, is one which seeks only to answer the questions: Does base stacking appreciably alter the vibrational spectra? Do differences in the identity of near neighbors produce different kinds of alterations in the spectra?

The first question has been treated in the work of Peticolas and co-workers [35,159], who have demonstrated that stacking of base residues in single-stranded polynucleotides, including poly(rA), gives rise to significant changes in the intensities of Raman lines associated with certain in-plane ring vibrations of the bases. A similar effect accompanies the stacking of monomeric nucleotides as discussed earlier [Sec. III.B.1.c.(2) and Fig. 21]. This phenomenon, termed *Raman hypochromism,* remains little understood, although a semiquantitative explanation in terms of the resonance Raman effect has been proposed [35]. Perturbations from base stacking have also been identified by Tsuboi [9] in IR spectra of polynucleotides.

As regards the second question posed above, several experimental approaches are feasible. One is to examine spectra of all possible dinucleotides, trinucleotides, etc., of known base sequence and for which data from other physicochemical studies confirm that the bases are stacked. For example, comparison of the spectra of the trinucleotides CpApC and UpApU should indicate whether the spectral bands of the A residue are perturbed differently when A is stacked between the different pyrimidines. A number of conclusions in this regard have already been reached from published data on dinucleoside monophosphates [161].

Another approach is to compare spectra of polynucleotides (or their complexes) which differ only in base sequence. A case in point is the duplexes poly(rA)·poly(rU) and poly(rA-rU)·poly(rA-rU). The former has been discussed above (Figs. 24 and 25). The latter is also a double-helical complex containing only AU pairs, but with alternating sequences of A and U residues in each strand. Thus each A residue is stacked between adjacent U residues and vice versa.

Significant differences between the IR spectra of poly(rA)·poly-(rU) (homopolymer complex) and poly(rA-rU)·poly(rA-rU) (copolymer complex) were first reported by Tsuboi [7]. These complexes have recently been studied in greater detail by Morikawa et al. [177], who have compared both the IR and Raman spectra and obtained evidence from a number of other physicochemical studies to confirm that the helix geometries are virtually identical except for base sequence. In Fig. 26, the IR spectra of the copolymer complex and its products of dissociation at 89°C are shown for comparison with Fig. 24. In Figs. 27 and 28 the corresponding Raman spectra are shown for comparison

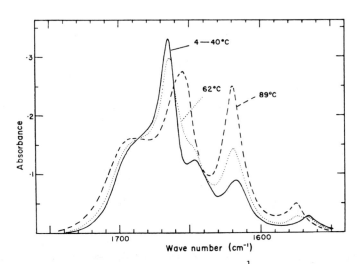

FIG. 26. IR spectra in the 1500 to 1750 cm^{-1} region of D_2O solutions of the double-helical complex poly(rA-rU)·poly(rA-rU) (4 to 40°C) and its products of dissociation (89°C). Note that the frequency increases from right to left along the abscissa, which is the reverse of Fig. 24. (From Ref. 177.)

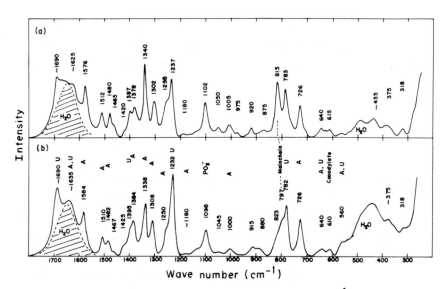

FIG. 27. Raman spectra in the region 300 to 1800 cm^{-1} of H$_2$O solutions of poly(rA-rU)·poly(rA-rU). (a) 32°C (double-helical complex); (b) 77°C (dissociated). (From Ref. 177.)

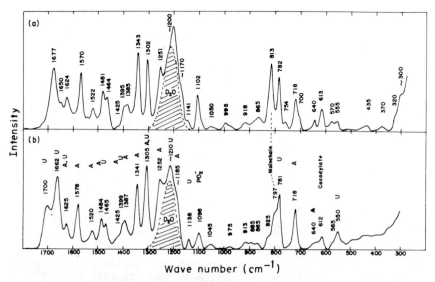

FIG. 28. Raman spectra in the region 300 to 1800 cm^{-1} of D$_2$O solutions of poly(rA-rU)·poly(rA-rU). (a) 32°C (double-helical complex); (b) 77°C (dissociated). (From Ref. 177.)

with Fig. 25. It will be noticed that the background of scattering by solvent (shaded) is not subtracted from Figs. 27 and 28. These data indicate the following: (1) at temperatures above the melting transitions, each dissociated complex gives the same IR spectrum in the double-bond region, which is the same as a superposition of the spectra of poly(rA) and poly(rU) at 85°C; (2) at temperatures above the melting transition, each dissociated complex gives the same Raman spectrum in the region 300 to 1800 cm^{-1}, which is only slightly different from the superposition of spectra of poly(rA) and poly(rU) at 85°C. Thus the Raman effect provides a more sensitive probe than IR spectroscopy for the residual base interactions which persist even at the elevated temperature; (3) at temperatures below the melting transition, the IR spectra of the two complexes exhibit major differences from one another; (4) at temperatures below the melting transition, the Raman spectra of the two complexes exhibit major differences from one another and from the corresponding IR spectra. These results demonstrate clearly that the frequencies and intensities in IR and Raman spectra are dependent upon the sequence of stacked bases in polynucleotide complexes and that the nature of the perturbations is different for the two kinds of spectra. Further discussion of the results in terms of the perturbations involved is given by Morikawa et al. [177].

(3) Conformational Dependence of Raman Scattering from Phosphate Group Vibrations. In the preceding sections, the influence of polynucleotide secondary structure on vibrations associated with the purine and pyrimidine residues has been discussed. We have seen that both base-pairing and base-stacking interactions may significantly alter the IR bands and Raman lines assigned to vibrations of the bases. We have also mentioned earlier (Sec. III.B.2.a) that the Raman line observed at ∿810 cm^{-1} for the phosphodiester (-O-P-O-) symmetric stretching vibration in polynucleotides is conformationally sensitive. We shall now discuss this point following the treatment given originally by Thomas and Hartman [172,186].

The initial assignment of the phosphodiester symmetric stretching frequency was made by Yu [178], who reported that Raman spectra
of poly(rA) at pH 7 showed a prominent line at 815 cm^{-1}. At pH 4,
when the adenine residues are protonated and a double-stranded complex [poly(rA$^+$)·poly(rA$^+$)] is formed, the -O-P-O- frequency occurs
at 824 cm^{-1}. Based on these results and the normal coordinate calculations of Shimanouchi et al. [6], Thomas [171] assigned the line
at 814 cm^{-1} in Raman spectra of aqueous rRNA to the same -O-P-O-
vibration and showed that it was sensitive in position and intensity
to changes in rRNA secondary structure. Subsequently, a number of
other investigators reported lines at or near this frequency in
Raman spectra of polynucleotides and nucleic acids (for reviews see
Refs. 4, 11, and 12). In Table 11 we have summarized the data on
phosphate group vibrations obtained from Raman spectra of both synthetic polynucleotides and nucleic acids. Several conclusions may
be drawn from these and similar data: (1) all frequencies are virtually unaffected by deuteration; (2) the symmetrical stretching
frequency of the PO$_2^-$ group (σ_2) invariably occurs at about 1100 cm^{-1}
with the same intrinsic intensity for ribo- and deoxyribopolymers,
regardless of conformational structure. In some cases, however, the
Raman line appears as a doublet (Table 11C); (3) the symmetrical
stretching vibration of the -O-P-O- group (σ_1) exhibits dependence
of frequency and intensity upon both the nature of the attached
sugar group (i.e., ribo vs. deoxyribo) and the polymer conformation;
(4) in the completely ordered ribopolymers (Table 11A), the intensity
ratio I_1/I_2 is at a maximum - approximately 1.64 - regardless of the
kind of secondary structure involved within the compounds listed;
(5) in completely disordered ribopolymers [e.g., poly(rA) and poly-
(rU) at high temperatures, Fig. 25], the intensity at 815 cm^{-1} and
the ratio I_1/I_2 fall to zero.

These results suggest that the value of I_1/I_2 in Raman spectra
of partly helical RNAs (Table 11, Section B) may be divided by 1.64
to obtain a quantitative estimate of the fraction of RNA phosphodiester groups which exist in the ordered configuration. Ordered

regions in this context include those containing either Watson-Crick pairs (AU and GC) or ordered single-stranded configurations such as occur in poly(rA) or poly(rC), since both types contribute equally to the value of I_1/I_2. Calculation of RNA secondary structure by this method shows that for most of the species listed in Table 11, Section B, approximately 85% of the residues are in ordered configurations [172,186].

It should be mentioned that the assignment of the frequency σ_1 to a symmetrical -O-P-O- stretching vibration does not preclude the possibility that the normal mode involved contains some coupling with adjacent groups, namely the C3', C5', or other furanosyl ring atoms. In fact some coupling with adjacent groups is required to explain the fact that the frequency in question here ($\sim$815 cm^{-1}) is substantially higher than the corresponding mode in dialkyl phosphate model compounds ($\sim$760 cm^{-1}) and phospholipids ($\sim$760 to 780 cm^{-1}). This view is also supported by recent IR data [187].

3. Nucleic Acids

Applications of IR and Raman spectroscopy in structural studies of nucleic acids have been extensively discussed in recent reviews [4,7,9,11,12] and in Parker's book [1]. Some additional observations on previous work will be mentioned here briefly.

a. Studies by Infrared Spectroscopy. IR absorption spectra of deuterated and nondeuterated films of DNA and RNA specimens are given in Fig. 29, where contributions from the nucleic acid residues and from adsorbed H_2O or D_2O are shown. Improved spectra may be obtained by use of polarized IR radiation on oriented samples under carefully controlled conditions of temperature and r.h. This is shown for the double-helical RNA of rice dwarf virus (RDV) in Fig. 30. Such spectra can be used to follow changes in nucleic acid conformation or to detect its molecular sites of adsorption of H_2O or D_2O as a function of temperature or r.h. [7,24]. A critical review of such studies on DNA and RNA samples has been given by Hartman et al. [11].

TABLE 11

Raman Frequencies and Relative Intensities of Phosphate Group Vibrations
in Polynucleotides and Nucleic Acids

Type of structure	H_2O Solutions			D_2O Solutions			Refs.
	σ_1(OPO)	σ_2(PO_2^-)	I_1/I_2	σ_1(OPO)	σ_2(PO_2^-)	I_1/I_2	
A. Completely ordered ribopolymers:							
single-stranded or Watson-Crick-paired							
Poly(rA)·poly(rU)	814	1101	1.63	812	1100	1.63	175
Poly(rA-rU)·poly(rA-rU)	815	1102	1.66	813	1102	1.65	177
Poly(rG)·poly(rC)	815	1101	1.63	813	1101	1.63	175
Poly(rI)·poly(rC)	811	1097	≈1.6	--	--	--	180
Poly(rA)	816	1098	≈1.6	814	1099	1.65	141,159,161,178
Poly(rC)	810	1101	≈1.6	810	1102	1.64	159,161,180
B. Partially ordered ribopolymers							
16-S rRNA (*E. coli* Q13)	814	1100	1.53	812	1100	1.56	33
23-S rRNA (*E. coli* Q13)	814	1100	≈1.4	812	1100	1.40	33
rRNA Fragments (*E. coli* MRE600)	814	1100	≈1.3	812	1100	≈1.4	171
rRNA Fragments (*E. coli* Q13)	814	1100	≈1.3	812	1100	≈1.3	--
tRNA^{fMet} (*E. coli* K12-MO7)	814	1100	1.38	812	1100	1.38	181
tRNA^{val} (*E. coli* B)	814	1100	1.36	812	1100	1.40	181

tRNA^{Val} (*E. coli* K12-M07)	814	1100	1.36	812	1100	≈1.4	--
tRNA^{Phe2} (*E. coli* K12-M07)	814	1100	1.40	812	1100	≈1.4	181
tRNA^{Arg} (*E. coli* K12-M07)	814	1100	1.38	812	1100	≈1.4	182
tRNA^{Glu} (*E. coli* K12-M07)	814	1100	1.38	812	1100	≈1.4	182
tRNA^{Phe} (Yeast)	814	1100	1.40	812	1100	1.40	189
R17-RNA	814	1100	1.44	812	1100	1.43	183
C. Other ribopolymers							
Poly(rG) (low ionic strength)	820	1087, 1100		823	1090, 1105		160
Poly(rG) (high ionic strength)	815	1092, 110?		815	1095, 1100		160
Poly(rI)	794, 822	1091		--	--		180
Poly(rU) (excess Mg^{2+})	814	1098		--	--		159
Poly(rA^{+})·poly(rA^{+})	824	1100		--	--		178
D. Completely ordered deoxyribopolymers							
Poly(dA)·poly(dT)	797, 840	1097		793, 841	1095		180
B-DNA (solution or gel)	787, 835	1094		792, 834	1095		184
A-DNA (gel)	807	1101		--	--		184
Poly(dG-dC)·poly(dG-dC) (low salt)	832	1094		--	--		185
Poly(dG-dC)·poly(dG-dC) (high salt)	814	1094		--	--		185

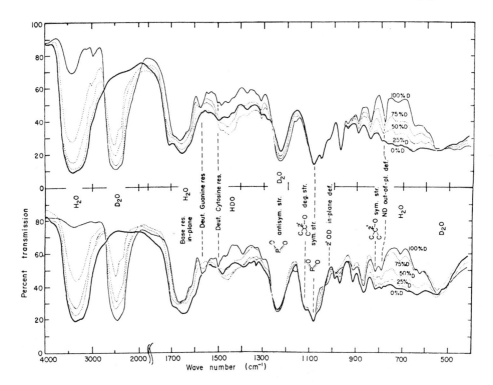

FIG. 29.   IR spectra of films of calf thymus DNA (upper) and unfractionated transfer RNA from yeast (lower) in air of 92% r.h.   The different curves correspond to nondeuterated, partly deuterated, and
completely deuterated samples as indicated.   (From Ref. 7.)

IR spectra of D_2O solutions of RNA in the double-bond region
(1450 to 1750 cm^{-1}) are also a source of structural information.
The absorption bands in this region have been assigned to stretching
vibrations of C=O, C=N, and C=C groups in the bases [6,7].   Each RNA
base (A, U, G, and C) gives rise to a unique pattern of bands which
is specifically altered by its pairing (hydrogen bonding) and stacking interactions within regions of RNA secondary structure.   As an
example of this effect we have cited above the IR spectra of poly(rA)·
poly(rU) and poly(rA-rU)·poly(rA-rU).   Similar data have also been
obtained on model structures containing G and C residues [176].   These
effects allow IR spectroscopy to be exploited as a determinant of RNA
secondary structure as follows.

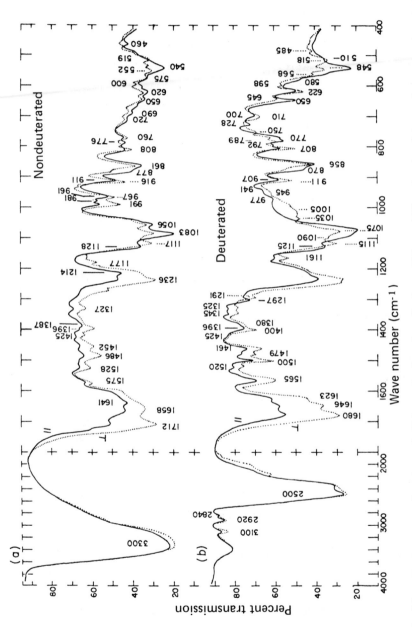

FIG. 30. IR spectra of oriented films of rice dwarf virus RNA at 75% r.h. in (a) nondeuterated and (b) deuterated forms. Solid curves are obtained for radiation polarized parallel to the fiber axis and broken curves for perpendicular polarization. (From Ref. 7.)

To obtain the percentages of RNA bases to a first approximation
that exist as AU and GC pairs in double-stranded regions and the percentages of unpaired bases in single-stranded regions, the IR spectrum of RNA is fitted to a synthesized spectrum obtained by summing
the appropriate quantitative reference spectra that correspond to
paired and unpaired bases [176]. For this purpose the spectra of
poly(rA)·poly(rU) and poly(rG)·poly(rC) have been used as reference
spectra for AU and GC pairs, respectively, and spectra of the 5'-
mononucleotides have been used as reference spectra for unpaired
bases. Implicit in such a procedure is the assumption that perturbations to the IR spectrum resulting from base stacking are small and
may be neglected in comparison to the much larger effects of base
pairing. This assumption will clearly introduce indeterminate errors
into the results, but the fact that use of this technique has con-
sistently given quantitative estimates of RNA secondary structure [12]
which are in full agreement with estimates available from other physicochemical investigations suggests that such errors are small. As
an example of this procedure, Fig. 31 compares the observed IR spectra
of 16-S and 23-S rRNA with those synthesized from the reference spectra of Thomas [176,179]. A summary of other applications has been
given by Hartman et al. [11].

    b. Studies by Raman Spectroscopy. The Raman spectra of H_2O
and D_2O solutions of a naturally occurring RNA are shown in Fig. 32.
Comparison with Fig. 31 reveals the advantages of the Raman technique
over IR spectroscopy for the study of aqueous RNA, and these advantages have been exploited in a number of investigations of secondary
structure in rRNA [33,171], tRNA [181,182,188-190], R17-RNA [183],
and MS2 RNA [87]. The Raman spectra of nucleic acids are sensitive
to the orientation of backbone residues, to hydrogen-bonding interactions of the bases (including base pairing), to base-stacking
interactions, and to binding of metal ions to phosphate or base substituents [33,171,182,189].

    (1) Unfractionated RNA. The first detailed Raman spectra of a
nucleic acid were obtained on unfractionated rRNA from *E. coli* ribo-

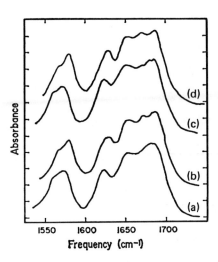

FIG. 31.   IR spectra in the region 1500 to 1750 cm^{-1} of 16-S and 23-S ribosomal RNA.   (a) Observed spectrum of 16-S rRNA; (b) synthesized spectrum for 16-S rRNA assuming 24% AU pairs, 36% GC pairs, 12.5% unpaired A, 9.5% unpaired U, 13.5% unpaired G, 4.5% unpaired C; (c) observed spectrum of 23S rRNA; (d) synthesized spectrum for 23-S rRNA assuming 25% AU, 33% GC, 13.0% A, 8.2% U, 16.3% G, 4.5% C. (From Ref. 179.)

somes [171].   These spectra are reproduced in Fig. 32 and the assignment of frequencies is given in Table 12.   The more intense lines in
the spectra result from vibrations of the bases (ring stretching).
The lines near 670, 720, and 780 cm^{-1} identify the G, A, and U + C
residues, respectively.   They are not appreciably shifted by deuteration and have been assigned to ring-stretching vibrations.   The intense lines in the interval 1200 to 1500 cm^{-1} are due to overlapping
ring vibrations of the bases.   The large deuteration effects indicate,
however, that these are strongly coupled to NH or ND deformation
vibrations of external amino and lactam groups [136].

     In the region 1500 to 1700 cm^{-1}, the lines are due to coupled
ring and double-bond stretching vibrations.   In D_2O solutions, those
at 1650 cm^{-1} and higher are mainly C=O stretching vibrations of U,
G, and C residues; in H_2O solution, they are coupled to NH deformations which accounts for the deuteration shifts.

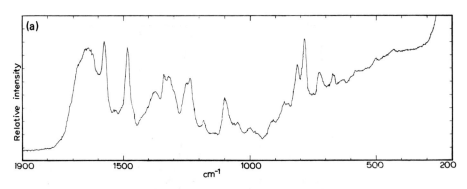

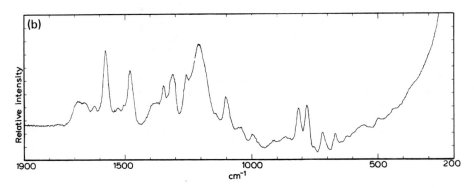

FIG. 32.   Raman spectra of unfractionated ribosomal RNA from *E. coli*
MRE 600.   (a) 35 mg/ml in H_2O solution at pH 7 and 32°C.   (b) 35 mg/ml
in H_2O solution at pD 7 and 32°C.   Scattering by solvent is not removed
from the spectra.   (From Ref. 171.)

The weak lines below 650 cm^{-1} are difficult to assign with certainty but may be due to out-of-plane ring deformations and to deformations of external C=O groups.   Contributions from the ribose-phosphate moiety may also be expected in this region.

The characteristic Raman frequencies of the phosphodiester group
at 814 and 1100 cm^{-1}, which have been discussed earlier, are also
evident in the RNA spectra.

When the ionic strength of solution is increased by the addition
of KCl, the Raman line at 1688 cm^{-1} increases in intensity at the
expense of the line at 1658 cm^{-1}, reflecting the formation of additional base pairs [171,179,191].   This effect is revealed in Fig. 33
(which may also be compared with Fig. 31 to indicate the differences

TABLE 12

Raman Frequencies and Assignments of Unfractionated rRNA from *E. coli* Ribosomes[a]

H_2O Solution	^2H_2O Solution	Nucleotide	Probable origin
435 (0)		U, C	Out-of-plane ring de-
500 (1)	498 (1)	G	formations; C=O defor-
580 (1B)	560 (1B)	A, U, G, C	mations, etc.
635 (0)	625 (0)	(A), U, C	
670 (2)	668 (2)	G	Ring stretching
710 (0S)	705 (0S)	C	
725 (3)	718 (3)	A	Ring stretching
	755 (0)		
786 (6)	780 (6)	U, C	Ring stretching
814 (5)	814 (6)	Phosphate	Symmetric stretching
867 (2)	860 (2B)	A, U, G, C	Ring stretching
918 (1)	915 (1)	Sugar, phosphate	-C-O- Stretching
975 (0)		Sugar, phosphate (?)	
	990 (1)	Sugar, phosphate (?)	
1003 (1)		A, U, C	
1049 (2)	1045 (1)	Sugar, phosphate	-C-O- Stretching
	1090 (S)	?	
1100 (5B)	1100 (4)	Phosphate	Symmetric stretching
	1140 (0)	A	
1185 (2)	1185 (1B)	A, U, G, C	Ring, external C-N
	1235 (S?)	A, C	stretching
1243 (6)		U, C	Ring stretching
1255 (5)	1257 (5)	A, C	Ring stretching
1300 (4S)		C	Ring stretching
	1310 (7)	A, U, C	Ring stretching
1320 (7)	1318 (6S)	A, G	Ring stretching
1340 (7)	1345 (3B)	A	Ring stretching
	1370 (3B)	A, G	
1380 (5B)		A, U, G	

TABLE 12 (continued)

H_2O Solution	^2H_2O Solution	Assignments	
		Nucleotide	Probable origin
	1390 (2S)	U	
1422 (2S)		A, G	Ring stretching
1460 (2S)	1460 (S)	U, C	CH Deformations
1484 (10)	1480 (8)	(A), G	Ring stretching
1510 (S?)	1503 (2S)	C	
	1560 (2S)	U	
1575 (8)	1578 (10)	A, G	Ring stretching
1620 (BS)	1622 (3)		Double bond stretching
1650 (BS)			vibrations of paired
	1658 (4B)		and unpaired bases;
1692 (4B)	1688 (4B)		mainly C=O stretching

[a]Vibrational frequencies are in cm^{-1}. Figures in parentheses denote relative intensities on the basis of 10 for the strongest Raman line in each spectrum. The symbols B and S denote broad and shoulder, respectively. Abbreviations: A, adenine; U, uracil; G, guanine; C, cytosine. Reproduced from Ref. 171, p. 420 (with alterations).

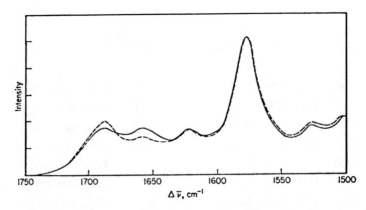

FIG. 33. Relative Raman intensities in the double-bond region for rRNA in D_2O (solid curve) and rRNA in D_2O containing 1 M KCl (broken curve). The spectra are normalized to give the same peak height at 1578 cm^{-1}. (From Ref. 171.)

in the double-bond region between Raman and IR spectra of identical RNA samples). In addition, the excess counterions cause a large increase in the intensity of Raman scattering at 814 cm^{-1}, reflecting the conversion of a large fraction of ribonucleotide residues of rRNA from disordered to ordered configurations. Subsequent studies have shown that the same conformational changes in rRNA may be produced by decreasing the solution temperature [33].

(2) Fractionated rRNA. Raman spectra of purified samples of 16-S rRNA in H_2O solution at 32, 60, and 80°C are shown in Fig. 34. Spectra of D_2O solutions and of solids have also been investigated, both for 16-S rRNA and 23-S rRNA [33,189]. The increase in temperature produces spectral changes which are qualitatively and quantitatively similar for each RNA, and which may be interpreted to reveal the changes in RNA secondary structure that have occurred [11].

A structurally informative region of the Raman spectrum (600 to 1000 cm^{-1}) for 16-S rRNA at different temperatures is shown in Fig. 35, after normalization of the intensities to an internal standard. Intensity alterations in the lines at 670, 780, and 814 cm^{-1} as a function of temperature are clearly evident, and these reflect substantial changes in secondary structure, namely, the unstacking of guanine and pyrimidines and the disordering of the backbone, respectively [11,33]. It may be noticed that while the 780-cm^{-1} line (ring frequency of the pyrimidines) is *hypo*chromic, the line at 670 cm^{-1} (ring frequency of G) is *hyper*chromic. Such *hyper*chromicity is frequently observed for lines of the guanine residue [11,161].

When native 16-S and 23-S rRNA molecules are compared with lower-molecular weight fragments produced from them [12], it is found that the latter contain fewer ordered secondary structures. This can be attributed to the presence of specific tertiary structures in the native molecules which confer a high degree of order on bases in single-stranded chain segments. Upon cleavage of the chains, the single-stranded segments are presumably free to assume more random configurations than are possible in the native molecules.

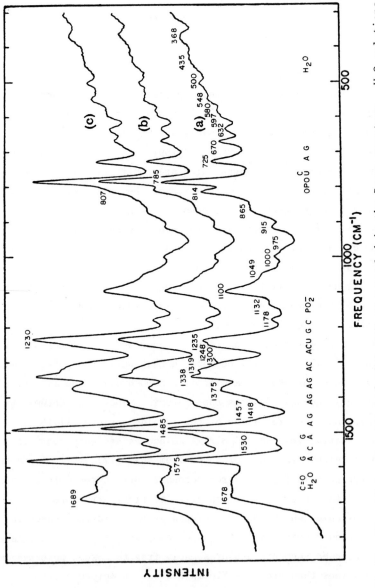

FIG. 34. Thermal denaturation of 16-S rRNA as revealed in the Raman spectrum. H_2O solutions at (a) 32°C, (b) 60°C, and (c) 80°C. (From Ref. 12.)

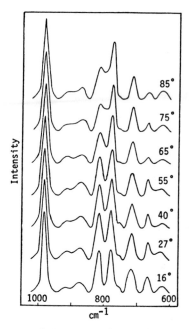

FIG. 35.   Raman spectra of D_2O solutions of 16-S rRNA (35 mg/ml) in the region 600 to 1000 cm^{-1} at different temperatures.   The spectra are normalized to the peak height of the 980-cm^{-1} line of SO_4^{2-} added as an internal standard.

(3) tRNA.   Raman spectra of several amino acid-specific tRNAs have been obtained (181,182,189,190].   Representative spectra are shown in Figs. 36 and 37.   tRNA spectra exhibit major differences from one another (and from spectra of rRNA), which can be attributed to differences in primary molecular structure (i.e., base composition).   For example, Fig. 38 reveals the different band patterns in the region 1150 to 1450 cm^{-1} exhibited by three different tRNAs. Nevertheless, all tRNAs that have been examined to date give, within the limits of experimental error, the same Raman intensity at 814 cm^{-1} (relative to that at 1100 cm^{-1}), as shown in Table 11B.   This fact indicates that each aqueous tRNA has the same percentage of its nucleotide residues in an ordered configuration and suggests a tertiary structure common to each tRNA, probably of the clover leaf type [172,181,182,186].

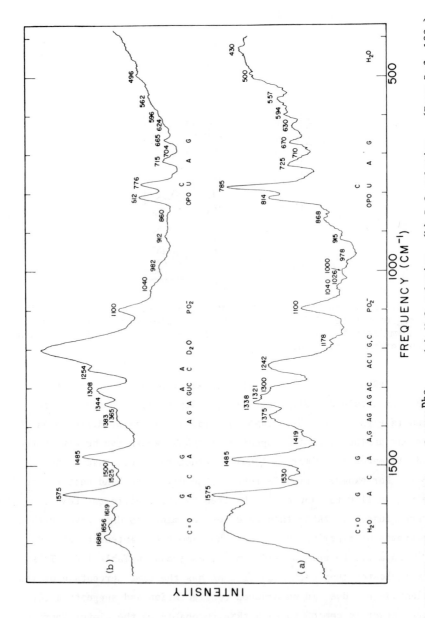

FIG. 36. Raman spectra of tRNA$_{Yeast}^{Phe}$. (a) H$_2$O solution, (b) D$_2$O solution. (From Ref. 189.)

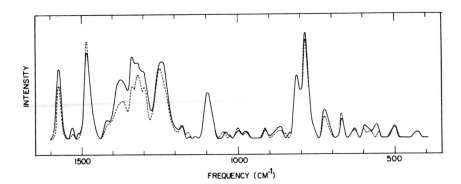

FIG. 37. Raman spectrum of tRNA$^{Phe}_{Yeast}$ (solid curve) compared with that of tRNA$^{Glu}_{E. coli}$ (broken curve). Redrawn from spectra of H_2O solutions. (From Ref. 189.)

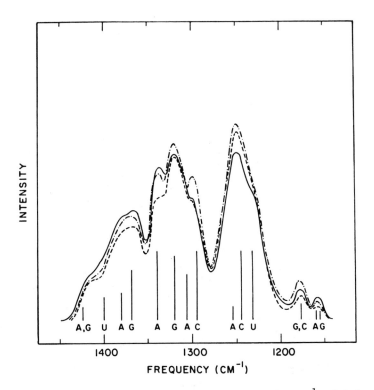

FIG. 38. Raman spectra in the region 1150 to 1450 cm^{-1} of H_2O solutions of tRNA$^{Glu}_{E. coli}$ (—·—·—), tRNA$^{Val}_{E. coli}$ (solid curve) and tRNA$^{fMet}_{E. coli}$ (broken curve). Vertical lines show the Raman scattering expected from ring vibrations of the bases. (From Ref. 182.)

The thermal denaturation of tRNA as observed in the Raman effect (Figs. 39 and 40) follows a pattern similar to that observed for rRNA. The denaturation is demonstrably reversible and occurs

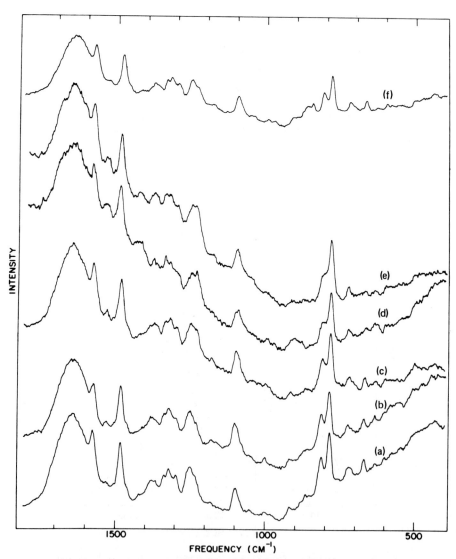

FIG. 39.   Raman spectra of H_2O solutions of $tRNA^{Glu}$ at different temperatures: (a) 32°C, (b) 50°C, (c) 65°C, (d) 80°C, (e) 90°C, and (f) 32°C after cooling from 90°C.   (From Ref. 182.)

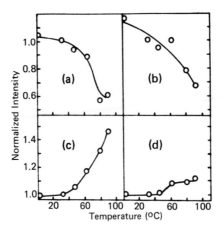

FIG. 40. Temperature dependence of Raman intensities at (a) 670 cm^{-1}, (b) 812 cm^{-1}, (c) 1578 cm^{-1}, and (d) 780 cm^{-1} in spectra of tRNA^{fMet} in D_2O solution. The hypochromicities [(c) and (d)] reflect the lesser stacking of purine and pyrimidine residues at elevated temperatures; the intensity decay at 812 cm^{-1} (b) reflects the disordering of the ribose-phosphate backbone; and the anomalous behavior of the 670-cm^{-1} intensity (a) has been attributed to the intrinsic hyperchromicity of this Raman line. (From Ref. 181.)

without breakage of the tRNA chains. From such spectra several conclusions have been drawn regarding the nature of changes in tRNA secondary structure which accompany heating [166,181,182,189].

A cautionary in the interpretation of Raman spectra of tRNAs is the contribution to the spectra from the various minor nucleosides. Figure 41 shows, for example, that dihydrouridine, which is present in nearly all tRNAs, will add to the Raman scattering at 720 cm^{-1}, otherwise due only to A residues. This contribution is small but significant [182].

The effect of Mg^{2+} on Raman spectra of tRNA has also been discussed [166,182,189].

(4) Aminoacylated tRNA. The effect of aminoacylation upon the Raman spectra of yeast tRNAs has been investigated recently [190]. A comparison of the spectra of charged and uncharged samples is given in Fig. 42, where it is seen that aminoacylation eliminates some of the Raman hypochromism in lines at 720 and 780 cm^{-1}, attributable to a reduction in stacking of adenine and pyrimidine residues.

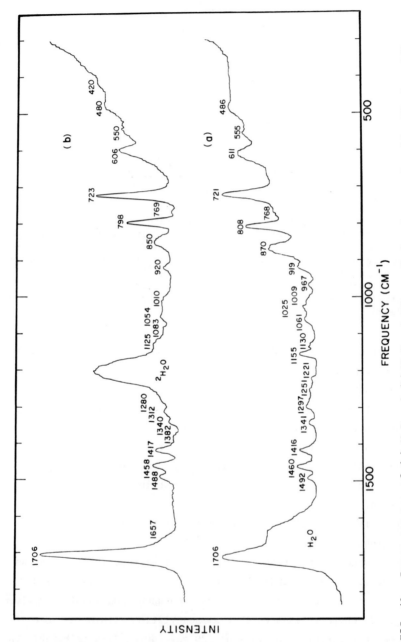

FIG. 41. Raman spectra of (a) H_2O and (b) D_2O solutions of the minor nucleoside dihydrouridine, present in tRNAs. (From Ref. 182.)

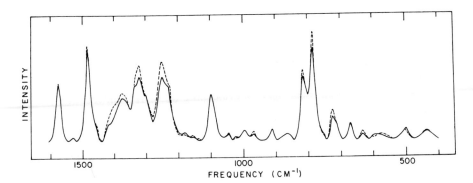

FIG. 42. Effect of aminoacylation upon the Raman spectrum of unfractionated tRNA from yeast. Solid curve, Non-aminoacylated tRNA; broken curve, aminoacylated tRNA. (From Ref. 190.)

(5) Viral RNA. The first Raman spectrum of a nucleoprotein was obtained on R17 phage [183], a small, nearly spherical virus which infects the bacterium *E. coli*. Each R17 particle consists of one molecule of RNA encapsulated by a coat of identical protein subunits. Hartman et al. [183] compared the Raman spectrum of the intact phage with the spectrum of R17 RNA, extracted from the phage, to reveal that only one significant difference exists between the Raman scattering of RNA in the phage and protein-free states. This involved the 1480-cm^{-1} line of guanine which was less intense than expected in the phage. Consequently, the intensity loss was attributed to a significantly different environment for guanine residues in protein-encapsulated RNA than in protein-free RNA.

A more detailed Raman study of another closely related viral RNA, from the bacteriophage MS2, has now been reported [87] and will be discussed in Sec. III.B.5.

(6) DNA. Raman spectra of calf thymus and salmon testes DNA were first published by Tobin [192], who investigated solids as well as aqueous solutions and gels under varying conditions of pH and pD. Many frequencies were detected but several of the assignments appear to be in conflict with earlier work on model compounds and with later work by other investigators on DNA.

Raman spectra of improved quality on H_2O and D_2O solutions of calf thymus DNA and on fibers drawn from the H_2O solutions have been published more recently [159,184]. In Fig. 43, the spectra of H_2O and D_2O solutions of DNA are shown for comparison with the RNA spectra of Fig. 32. The frequency assignments of Small and Peticolas [159] are given in Table 13. The major differences between spectra of RNA and DNA are due to the replacements of uracil by thymine and of ribose by deoxyribose. An important characteristic of the Raman spectra of DNA in aqueous solution is the absence of a line near 814 cm^{-1}, present in RNA and assigned to the symmetric -O-P-O-stretching vibration of the riboester linkages in an ordered configuration. This fact alone demonstrates that the backbone conformation in aqueous double-helical DNA differs from that which occurs in aqueous double-helical polyribonucleotides and RNA (Table 11).

The Raman spectra of fibers of DNA at different r.h. values are shown in Fig. 44. At 75% r.h., when DNA is presumably in the A form [193], a prominent Raman line appears at 807 cm^{-1}. Its proximity to the line at 814 cm^{-1} in ribopolymers suggests that in A-DNA the backbone conformation is similar to that which occurs in the aqueous RNAs. Likewise, the similarities in phosphate group frequencies of aqueous DNA and fibrous DNA at 98% r.h. (i.e., B-DNA) suggest for these a common backbone structure [184].

Thermal denaturation of DNA has also been studied by Raman spectroscopy. Two groups [194,195] have independently reported that the Raman spectrum is sufficiently sensitive to monitor changes in DNA secondary structure (so-called "premelting") before the onset of true denaturation (cooperative melting). The study of Rimai et al. [194] demonstrates that in the premelting region of calf thymus DNA (20 to 60°C), the Raman spectrum, like the near-u.v. absorption spectrum, exhibits a number of dramatic intensity changes, but these can be masked by the fact that some of the Raman lines in question are relatively weak and partially overlapping of one another, which could account for the failure of other workers [35,159,195] to detect these features. An important conclusion derived from these results

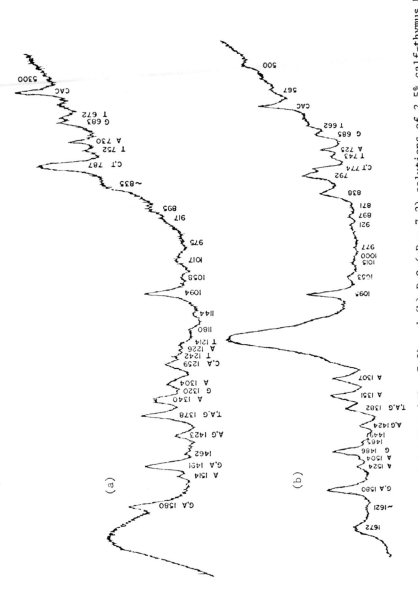

FIG. 43. Raman spectra of (a) H₂O (pH = 7.2) and (b) D₂O (pD = 7.2) solutions of 2.5% calf-thymus DNA. (From Ref. 12.)

TABLE 13

Raman Frequencies and Assignments for Calf Thymus DNA[a]

Frequencies (cm^{-1})		Assignments[b]
H$_2$O Solution	D$_2$O Solution	
	500	Deoxyribose-phosphate
	567	Deoxyribose
672	662	T
683	685	G
730	725	A
752	743	T
	774	C, T
787		O——P——O Diester symmetric stretch overlapping C,T
	792	O——P——O Diester symmetric stretch
∿835	838	Deoxyribose-phosphate
	871	Deoxyribose-phosphate
895	897	Deoxyribose-phosphate
917	921	Deoxyribose
975	977	Deoxyribose
1017	1015	C——O Stretch
1058	1053	C——O Stretch
1094	1095	O···P···O$^-$ Symmetric stretch
1144		Deoxyribose-phosphate
1180		Base external C——N stretch
1214		T
1226		A
1242		T
1259		C, A
1304	1307	A
1320		G
1340	1351	A
1378	1382	T, A, G

TABLE 13 (continued)

Frequencies (cm^{-1})		Assignments[b]
H$_2$O Solution	D$_2$O Solution	
1423	1424	A, G
1448	1449	Deoxyribose
1462	1465	Deoxyribose
1491	1486	G, A
	1504	A
1514	1524	A
1534		G, C
1580	1580	G, A
	∿1621	
	1672	C=O Stretch

[a]Reproduced from Ref. 159, p. 1399.

[b]T, C, A, and G indicate vibrations characteristic of the thymine, cytosine, adenine, and guanine bases, respectively, listed in order of their relative contributions with the largest contribution first. Deoxyribose-phosphate indicates probable origin is in the deoxyribose-phosphate chain but cannot be readily assigned specifically to deoxyribose or phosphate.

is that the Raman-scattering cross sections (i.e., intensities) for base vibrations can be dependent upon two or more electronic states determining the Raman tensor. Therefore, simplified treatments which attempt to correlate the Raman intensities with one electronic state only (e.g., the state giving rise to absorption near 260 nm in u.v. spectra of polynucleotides) may lead to erroneous conclusions. The results of Rimai et al. [194] also show that hyperchromic effects detected previously in RNA [33] and polyribonucleotides [175], as well as in guanine ribonucleotides [159], are not peculiar to the ribopolymers but are more likely a general feature of Raman scattering of nucleic acids due to the fact that many different electronic states in the near-u.v. determine the scattering cross sections for the base vibrations.

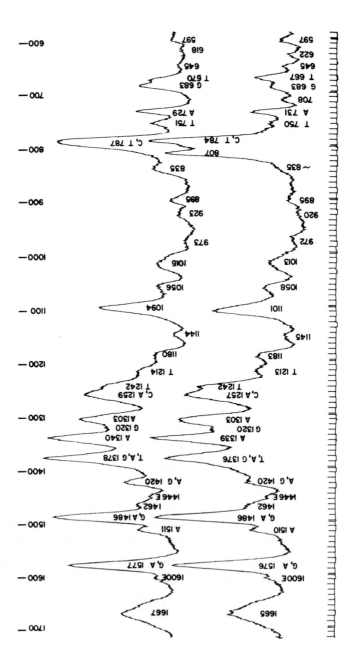

FIG. 44. Raman spectra of fibers of calf-thymus DNA at 98% (upper) and 75% (lower) r.h. (From Ref. 184.)

4. *Special Problems in Vibrational Spectra of Nucleic Acids and Derivatives*

a. *Deuterium Exchange of Hydrogen at the 8-C Position of Purines.* Investigations by nmr spectroscopy [196,197] have revealed that exchange of hydrogen and deuterium at the 8-C ring position of purines proceeds rather easily at elevated temperatures. By means of tritium exchange, the reaction kinetics in adenine and guanine derivatives [137] and in various polynucleotides [198,199] have been determined. The effects of deuterium exchange of hydrogen on the Raman spectrum of purines was first demonstrated for the case of inosine [139] and has recently been followed with detailed studies on a number of adenine derivatives [141,200].

The 8-CH exchange reaction is of importance here for two reasons. First, substitution of deuterium for hydrogen at the 8-C position of purines is expected to occur in D_2O solutions of purine-containing nucleotides, polynucleotides, and nucleic acids, especially at elevated temperatures. Such exchanges may also occur in films or fibers placed in a D_2O atmosphere for prolonged periods at ambient temperatures. When exchange takes place the IR and/or Raman spectra may be significantly altered by virtue of the dependence of vibrational frequencies on nuclear mass. Spectral changes resulting only from exchange must therefore be well understood before the spectra can be interpreted in terms of other structural or conformational changes. In practice, complications resulting from exchange may usually be avoided by assuring that all such exchanges are completed before the sample is examined spectroscopically. For the 8-CH group, exchange can be completed by prolonged heating in D_2O solution. However, for fragile biopolymers, such as high-molecular weight RNA, heat treatment may result in degradation. When 8-CH exchange cannot be accomplished safely without damaging the polymer, the spectral effects of exchange can at least be recognized by the fact that they are not reversible [139].

Second, the vibrational spectrum provides a convenient method for investigating the kinetics of isotope exchange reactions [Sec. I.B.2.d). A preliminary Raman study of the exchange at 8-C in inosine

was reported previously without analysis of the exchange kinetics
[139]. It is now demonstrated [200] that accurate quantitative study
of the exchange reaction is easily carried out by Raman spectroscopy,
and the results obtained on several adenine derivatives permit subtle
differences in their conformational structures to be identified [141].
Figures 45 and 46 demonstrate the changes which occur in the Raman

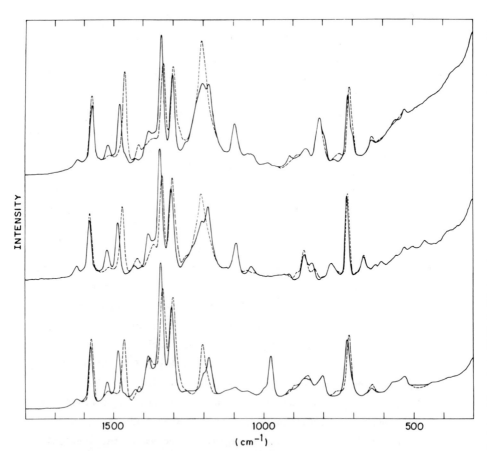

FIG. 45. Raman spectra in the region below 2000 cm^{-1} of D$_2$O solutions of Ado-5'-P (lower), Ado-3':5'-P (center), and Poly(rA) (upper).
Solid curve, 8C-H form of adenine ring; broken curve, 8C-D form of
adenine ring. (From Ref. 141.)

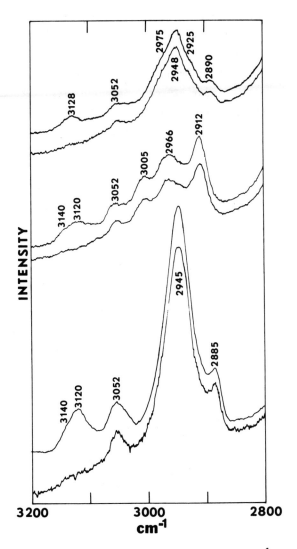

FIG. 46. Raman spectra in the region 2800 to 3200 cm^{-1} of D_2O solutions of Ado-5'-P (lower), Ado-3':5'-P (center), and Poly(rA) (upper). For each pair of curves the upper and lower members correspond to the 8-CH and 8-CD forms, respectively, of the adenine ring. (From Ref. 141.)

spectra of Ado-5'-P, Ado-3':5'-P, and poly(rA) upon 8-CH exchange by
deuterium. After exchange is completed, the line at 1485 cm^{-1} is
replaced in each case by a line at 1465 cm^{-1}. Other large changes
in the spectrum are also evident from Fig. 45. It is clear, however,
that the Raman lines near 720 cm^{-1} (adenine ring), 980 cm^{-1} (PO_3^{-2}
group), and 1095 cm^{-1} (PO_2^{-} group) are unaffected by exchange and
thus provide a means of internal standardization of other intensities
with respect to the exchange reaction. From these and other spectra,
sufficient data are obtained to determine the pseudo-first order
rate constant (k_{ψ}) governing each exchange reaction. The temperature
dependence of k_{ψ} in turn provides the Arrhenius parameters governing
exchange, as shown in Table 14. These results reveal an acceleration
of exchange in Ado-3':5'-P relative to Ado-5'-P at low temperature
and a retardation of exchange in poly(rA) at all temperatures which
are explained in terms of molecular conformation [141,201].

    b. *Effects of Base Sequence on Vibrational Frequencies and
Intensities.* The perturbations to IR and Raman spectra which result
from base-stacking interactions in polynucleotides and nucleic acids

TABLE 14

Arrhenius Parameters Governing 8-CH

Exchange in Adenine Nucleotides*

Sample (temperature)		Activation energy, E_a (kcal/mole)	Frequency factor, A (hr^{-1})
Ado-5'-P	(20-80°C)	24.2 ± 0.6	2.3×10^{14}
Ado-3':5'-P	(<50°C)	17.7 ± 0.1	9.6×10^{9}
	(>50°C)	23.5 ± 0.6	8.3×10^{13}
	(30-80°C)	21.5 ± 1.5	4.3×10^{12}
Poly(rA)	(<60°C)	27.7 ± 0.4	1.8×10^{16}
	(>60°C)	22.0 ± 0.1	3.2×10^{12}
	(30-90°C)	25.3 ± 1.4	3.5×10^{14}

*For further details consult Ref. 141. (This table was reproduced
 with alterations from Ref. 141, Table 3.)

have been discussed in the preceding sections. From what has been said, it is clear that further study of model compounds is necessary to understand the origins of these perturbations and their contributions to spectra of nucleic acids. First steps in this direction have been taken in the studies by Morikawa et al. [178] and Prescott et al. [161].

   c. Normal Coordinate Calculations. Nucleic acids are complex molecules, each nucleotide containing 30 or so atoms. The vibrational spectra of the nucleotide units are exceedingly complex, and reliable assignment of each IR band or Raman line is not possible from studies of model compounds alone. Calculation of the normal modes of vibration, by use of force constants transferred from other simpler molecules, may facilitate assignments and enhance the usefulness of the vibrational spectra as a probe of nucleic acid structures. Thus, the assignment of phosphate group frequencies in nucleic acids has been clearly advantaged by the normal coordinate calculations of Shimanouchi et al. [6]. Extension of the calculations to vibrations of the purine and pyrimidine bases has been undertaken recently [10, 89,158]. The reliability of these results should be further increased when additional data are obtained from isotopically substituted derivatives.

   d. Preresonance Effects in the Raman Spectra. Tsuboi et al. [158,202] have compared Raman spectra of uracil derivatives excited with argon (4880 Å) and helium-neon (6328 Å) lasers. When 4880 Å excitation is employed, the Raman intensities for uracil ring vibrations (relative to the intensities for phosphate group vibrations) are considerably greater than when 6328 Å excitation is used. This fact is ascribed to the so-called preresonance Raman effect, i.e., the greater proximity of the 4880-Å line to the electronic absorption band of the uracil residue at 2600 Å enhances the intensity of Raman lines associated with uracil ring vibrations that derive their Raman scattering cross sections in whole or in part from the 2600-Å electronic state. On the other hand, the Raman lines of the phosphate residue, which has no appreciable electronic absorption down to 1800 Å, are virtually unaffected by the difference in wavelength of excitation.

The quantitative measurement of Raman line intensities under conditions which favor preresonance may therefore be helpful in assigning the frequencies to vibrations of specific functional groups as well as in revealing information about the excited electronic states. Accordingly several investigators are now undertaking close examination of the Raman intensities of nucleic acid base vibrations under conditions of excitation which approach resonance with the 2600-Å band. These studies [203-205] have revealed that different Raman lines derive their intensities from different electronic states, which accounts for the fact that both hypochromic and hyperchromic effects can be observed simultaneously in the Raman spectra of nucleic acid biopolymers [33,175] when secondary structures are eliminated. The value of preresonance and resonance Raman spectra of nucleic acids should be greatly enhanced when tunable lasers in the u.v. region become generally available.

5. *Nucleic Acid-protein Systems*

a. Nucleoproteins. Applications of IR spectroscopy in the study of nucleoprotamine and nucleohistone have been reviewed by Parker [1]. In a more recent study, Cotter and Gratzer [206] compared IR spectra of *E. coli* ribosomes with those of rRNA and protein extracted from the ribosomes to reveal similarities in rRNA structure in the protein-free and protein-bound states. The amount of rRNA secondary structure is determined from the IR intensities in the double-bond region of spectra of D_2O solutions by the same procedure [176], discussed earlier (Sec. III.B.3.a).

Prescott et al. [258] have recently succeeded in obtaining Raman spectra of complexes of DNA and poly(rA) with poly-L-lysine, which reveal the extent of conformational changes in both nucleic acid and polypeptide backbones upon complex formation.

b. Viruses. In recent years, a heightened interest in the mechanism and control of viral infection has promoted an intensification of research on the structures and interactions of nucleic acid and protein components of viruses [207]. Raman spectroscopy appears

to be a powerful method for investigating virus assembly and determining details of the virion structure in aqueous media [87,88,183].

In a preliminary study of the bacteriophage R17, Hartman et al.
[183] showed that the R17 virion, which by weight contains approximately 30% RNA and 70% protein, yields a Raman spectrum in which
lines of both the RNA and protein components are clearly resolved
from one another. Their data are tabulated in Table 15, with assignments based upon their earlier studies of model compounds. A more
detailed study of the bacteriophage MS2, closely related to R17 both
in function and chemical composition, has now been carried out [87].

Fig. 47(a) shows the Raman spectra of both H_2O and D_2O solutions
of MS2 phage. The prominent lines are assigned to RNA and coat protein subgroups in the same way as for R17 phage (Table 15). Lines
near 670 (G), 720 (A), 780 (U + C), and 815 (OPO) are the most useful
in terms of studying nucleotide conformations, while lines at 1663
(amide I), 1250 (amide III), and 1236 (amide III) are the conformationally sensitive contributions from coat protein. These features
reveal immediately a highly ordered secondary structure for the encapsulated RNA molecule and a predominantly β-sheet and random chain
conformation for coat protein backbones. The spectra of phage are
not much different from a sum of spectra recorded separately on the
coat protein and RNA; however, the temperature and ionic strength
dependence of Raman scattering from the phage differ strikingly from
that of RNA and protein components. Thomas et al. [87] have used
these results to reach a number of conclusions about the structural
stability of the virion and its nucleic acid and protein components.
Of particular interest is the finding that the coat protein confers
a high degree of stability to the otherwise fragile secondary structure of RNA. Moreover, this stabilizing effect is more pronounced
in solutions of high ionic strength, from which it can be inferred
that hydrophobic interactions are responsible, at least in part, for
the proper assembly of the virion. Raman spectra of the MS2 capsids
(i.e., viral particles from which all RNA has been extruded) reveal
a highly stable structure with respect to changes in either temperature or ionic strength. Thus the capsid does not depend upon protein-
RNA interactions for its stability.

TABLE 15

Raman Frequencies and Assignments of R17 Virus

and R17 RNA[*]

R17 RNA		R17 Virus		
Frequency	Assignment	Frequency	Assignments	
			RNA	Protein
		∿350(0)	r, C	
		385(0)	C	
435(1)	r	435(1)	r	?
502(0)	G,C	500(1)	G,C	Disulfide
580(1B)	C,G	575(1)	C,G	Trp
635(1)	r	637(1)	r	Disulfide
670(2)	G	670(2)	G	
710(S)	A	710(S)	A	
725(3)	A	722(3)	A	
755(0)	C	760(3)		Trp
787(10)	C,U	787(8)	C,U	
815(7)	P	815(6)	P	
		830(S)		Tyr
860(0)	r	865(3)	r	Trp
880(0)	r	885(S)	r	Trp
915(1)	r			
975(0)	r	930(1)		Trp
1001(1)	A,U,C	1005(7)		Phe
		1015(S)		Trp
1045(1)	r	1035(1)	r	Phe
		1085(S)		Phe
1100(5)	P	1100(5)	P	
		1125(1)		?
1160(0)	r	1162(0)	r	
1180(1)	C,G,A	1180(0)	C,G,A	
		1210(S)		Tyr, Phe
1238(6)	U,C	1238(10)	U,C	Am III
1250(6)	C,U	1250(8)	C,U	Am III
		1268(S)		Am III

TABLE 15 (continued)

R17 RNA		R17 Virus		
Frequency	Assignment	Frequency	Assignments	
			RNA	Protein
1303(4)	A,C	1300(S)	A,C	
1320(6)	G	1320(S)	G	
1342(6)	A	1343(6)	A	Trp
		1365(2B)	G	Trp
1380(4B)	G,U,A	1400(1)	U,A	
1462(0)	C-H def	1450(S) 1462(4)	C-H def	
1482(10)	G,A	1480(5)	G,A	
1535(0)	G	1550(S)		Trp
1575(7)	G,A	1575(5)	G,A	Trp
		1600(1)		?
1620(B)	U	1620(2)	U	Trp,Tyr,Phe
∿1650(B) ∿1685(B)	U,G,C	∿1665(B)	U,G,C	Am I
2890(1) 2955(6) 2980(3)	Aliphatic C-H str	2880(10B) 2942(30B) 2975(15B)	Aliphatic C-H str	

*Frequencies in cm^{-1} are accurate to ± 3 cm^{-1} for sharp lines and ± 5 cm^{-1} for weak or broad lines. Relative intensities, in parentheses, are based upon an arbitrary intensity of 10 given to the strongest line below 1600 cm^{-1} in each spectrum. Details of assignments are given in Ref. 183. Abbreviations: A, adenine; U, uracil; G, guanine; C, cytosine; r, ribose; P, phosphate; trp, tryptophan; tyr, tyrosine; phe, phenylalanine; am, amide; str, stretching; def, deformation; S, shoulder; and B, broad. Reproduced from Ref. 183, p. 945.

DNA viruses have also been investigated by Raman spectroscopy [88]. The filamentous phages Pfl and fd, which contain only 12% DNA by weight, yield Raman spectra which are dominated by vibrations of the coat protein [Fig. 47(b)]. The data indicate that the coat protein molecules are α-helical and that the encapsulated DNA does not exist in the A conformation.

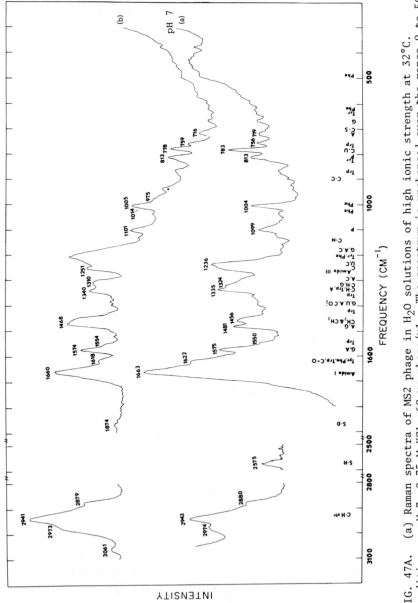

FIG. 47A. (a) Raman spectra of MS2 phage in H₂O solutions of high ionic strength at 32°C. Conditions: pH 7, 0.75 M KCl, 60 μg phage/μl. The spectrum is unchanged over the range 0 to 50°C. (b) Raman spectra of MS2 phage in D₂O solutions of high ionic strength at 32°C. Conditions: pD 7, 0.75 M KCl, 79 μg phage/μl. The spectrum is unchanged over the range 0 to 50°C. (From Ref. 87.)

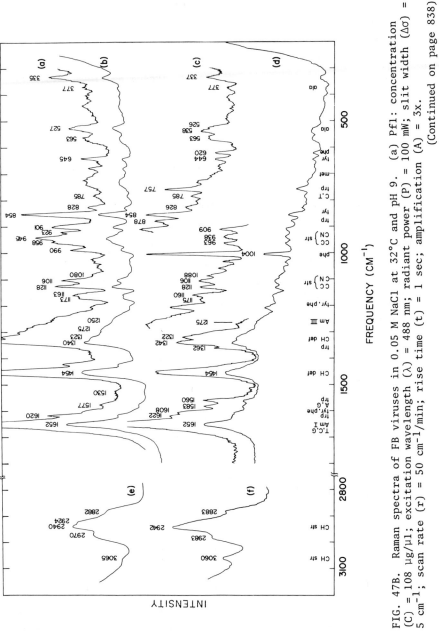

FIG. 47B. Raman spectra of FB viruses in 0.05 M NaCl at 32°C and pH 9. (a) Pfl: concentration (C) = 108 μg/μl; excitation wavelength (λ) = 488 nm; radiant power (P) = 100 mW; slit width (Δσ) = 5 cm⁻¹; scan rate (r) = 50 cm⁻¹/min; rise time (t) = 1 sec; amplification (A) = 3x.

(Continued on page 838)

Thus Raman spectra of high resolution are obtainable from biologically active viruses, and the spectra contain considerable structural information about viral nucleic acid and protein components.
It is safe to predict that the Raman effect will also find use as a
structural probe of other nucleoproteins, including ribosomes, chro-
mosomes, and more complex viruses.

C. Lipids and Membranes

Proteins and nucleic acids, discussed in the two preceding sections (III.A and B, respectively), can be considered *simple* biological
macromolecules in the sense that biological activity is contained
within single (or at most a few) molecular units of the same or similar kind. Biological membranes, however, are assemblages of many
different kinds of molecular units, usually including lipids, proteins,
and saccharides, and occasionally, nucleotides and visual pigments.
A minimally functional membrane or model membrane, therefore, is a
complex biological organelle. Accordingly, the vibrational spectrum
of a membrane is exceedingly complicated, typically containing many
overlapping bands or lines from different molecular constituents, and
not easily interpreted in terms of the structures or interactions of
the component molecules.

In addition to difficulties in the analysis of spectra, it should
be noted that the structure and functions of a membrane are determined
not only by the structures of the composite molecules, but also by
their mode of packing in the functional unit. It is well to keep in
mind the complexities of membranes when attempting to apply IR and
Raman spectroscopy to them.

[FIG. 47B (continued)] (b) Pf1: $\Delta\sigma = 10$ cm^{-1}; t = 3 sec; A = 1x; other
conditions as in (a). (c) fd: C = 147 µg/µl; λ = 514.5 nm; P = 100 mW;
$\Delta\sigma = 5$ cm^{-1}; r = 50 cm^{-1}/min; t = 1 sec; A = 3x. (d) fd: $\Delta\sigma = 9$ cm^{-1};
t = 3 sec; A = 1x; other conditions as in (c). (e) Pf1: A = 1/3x;
other conditions as in (b). (f) fd: A - 1/3x; other conditions as in
(d). Frequencies of prominent lines are given in cm^{-1} units and assignments to molecular subgroups are denoted by standard abbreviations.
Abbreviations: str, stretching; def, deformation; CH, carbon-hydrogen
bond; CC, carbon-carbon bond; CN, carbon-nitrogen bond; A, T, C, and G,
adenine, thymine, cytosine, and guanine; ala, alanine; met, methionine;
phe, phenylalanine; trp, tryptophan; tyr, tyrosine; am, amide. (From
Ref. 88.)

Among the many constituents of membranes, lipids--especially
phospholipids--are the most important in providing structural order
and specific morphological properties. Ordinarily, a membrane must
function to (1) separate two media of differing chemical properties
(the "inside" and "outside" surfaces of the membrane), (2) direct the
transport of ions or other substances from one medium to the other,
and (3) provide a matrix in which enzyme-catalyzed reactions can take
place. It is in the first of these roles that lipids are believed
to play the major part. Lipids themselves can form micellar structures in aqueous environments, and considerable effort has been devoted in recent years to studying the mechanism of micelle formation
by lipids. The interactions between lipid molecules or between lipid
and solvent molecules have also been widely studied. In membrane
research, IR and Raman spectra have been used mainly towards these
ends.

In the discussion which follows, several examples are considered
which illustrate the type of data obtained and the conclusions derived
therefrom. We also discuss briefly the principles which govern vibrational spectra of medium- and long-chain aliphatic hydrocarbons as
are found in lipids and lipid-like molecules.

1. *Fatty Acids*

a. *Infrared Spectra.* The IR spectra of fatty acids are relatively
simple for the liquid and solution states. However, in solid state
spectra a progression of bands is often observed. This phenomenon
is most typical for (1) the methylene (CH_2) wagging or twisting vibrations which give a progression of bands spaced fairly evenly between 1150 and 1350 cm^{-1}, (2) the CH_2 rocking vibrations which give
a progression between 700 and 1050 cm^{-1}, and (3) the skeletal vibrations which occur below 600 cm^{-1}. To exemplify the progressions
which arise from the CH_2 motions, we show in Fig. 48 the absorption
spectrum of methyl stearate over most of the mid-IR region.

In general, for a free-end polymer with n atoms per translational
repeat unit and M repeat units, there are 3Mn normal modes of vibration (including the six zero frequency modes). These vibrations are

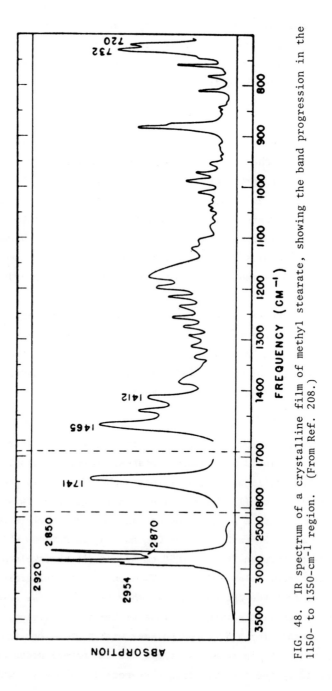

FIG. 48. IR spectrum of a crystalline film of methyl stearate, showing the band progression in the 1150- to 1350-cm⁻¹ region. (From Ref. 208.)

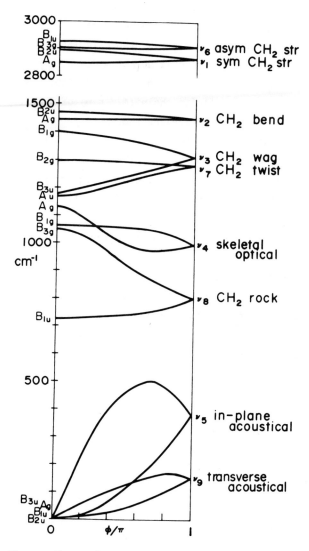

FIG. 49. Phonon dispersion curves of a polyethylene chain in the zigzag (all-trans) conformation. (From Ref. 209.)

distributed in 3n frequency branches at phase angles ϕ_k, where

$$\phi_k = \frac{k\pi}{M} \quad k = 0, 1, 2,\ldots, M - 1 \tag{11}$$

The vibrational frequency for a given normal mode may be plotted as

a function of ϕ_k/π to give what is usually referred to as a phonon
dispersion curve [260]. In Fig. 49, the calculated phonon dispersion curves are shown for several normal modes of a polyethylene
chain in the all-trans (zig-zag) conformation [209]. (See also Chap.
1.) Vibrations of an infinite chain (ϕ/π = 0) of symmetry species
B_{1u}, B_{2u}, and B_{3u} are IR-active and those of species A_g, B_{1g}, B_{2g},
and B_{3g} are Raman-active. The calculated band positions for the IR-
active CH_2 wagging vibrations of several hydrocarbon chains are shown
in Fig. 50 [210]. Data such as these are frequently helpful in
assigning the band progressions observed in spectra of long-chain
fatty acids and their derivatives.

Absorption spectra in the mid-IR region are found to be relatively insensitive to trans-gauche isomerizations in compounds containing saturated hydrocarbon chains. Spectra in the far-IR region,
while being more sensitive to such structural changes, are more difficult to obtain for experimental reasons associated with the use of
the far-IR, although recent advances in instrumentation are helping
to overcome these difficulties. The Raman spectrum thus provides a
more practical means of investigating conformationally sensitive low
energy frequencies of the hydrocarbon chain in saturated fatty acids.

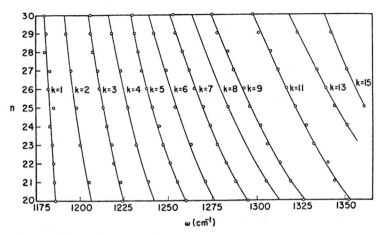

FIG. 50. Array of CH_2 wagging frequencies for polyethylene chains of
varying length (number of repeat units, M) as a function of k [see
text, Eq. (11)]. (From Ref. 210.)

In compounds containing unsaturated hydrocarbon chains, cis and trans geometries with respect to C=C bonds are possible. The out-of-plane bending vibration of the olefinic C-H group gives rise to prominent IR bands at 910 and 970 cm^{-1} in cis and trans forms, respectively. Thus the IR data may be profitably used to distinguish between cis and trans isomers of unsaturated fatty acids [211].

   b. Raman Spectra. Raman spectra of saturated fatty acids with hydrocarbon chains in the all-trans conformation give characteristic lines at 1064 and 1130 cm^{-1}, while those with hydrocarbon chains in the gauche form give a characteristic line at 1089 cm^{-1}. It thus appears that these Raman frequencies are useful in distinguishing between the two conformers [212].

   Long-chain hydrocarbons also give rise to two skeletal vibrations (known as ν_4 and ν_5) [210], which are intense in the Raman effect and potentially useful for conformational studies. The ν_5 (k = 1) vibration, or so-called accordion mode, and its overtones have been assigned by Schaufele and Shimanouchi [223] to the series of lines below 300 cm^{-1} in the Raman spectrum of single crystals of low-molecular-weight polyethylene, $C_{36}H_{74}$ [Fig. 51(a)]. The frequencies fit the formula

$$\nu = \frac{am}{L} \tag{12}$$

where a is a factor depending upon mass and force constant, m an integer, and L the hydrocarbon chain length. Conversely, Eq. (12) may be applied to determine the hydrocarbon chain length when frequencies are reliably assigned in spectra of fatty acids [209].

   While the saturated fatty acids give a single Raman line for ν_5, corresponding to k = 1 [Eq. (11)] and m = 1 [Eq. (12)], unsaturated fatty acids give more complicated spectra in this frequency region. Possible structural correlations have been discussed by Lippert and Peticolas [209].

   Very recent developments show that accordion-like vibrations occur even for aqueous fatty acids with fairly short chain lengths [216,217]. Aqueous solutions of the sodium-n-alkyl sulfates and potassium-n-alkyl carboxylates (including members for which n = 5 or 6) show a definite series of Raman lines below 800 cm^{-1} [Fig. 51(b)]. From the positions and intensities of these lines it can be concluded that the hydrocarbon chains assume the all-trans configuration. Randomness in the conformation of the n-alkyl hydrocarbon skeleton of these surfactant molecules also is found to

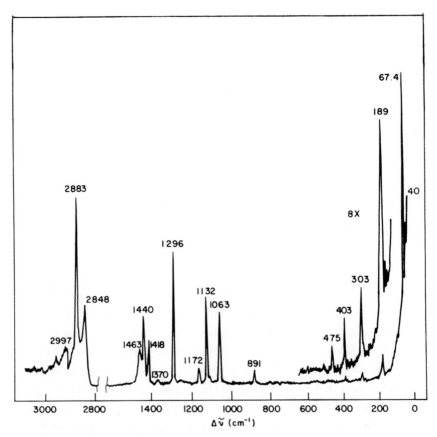

FIG. 51(a). Raman spectrum of the crystalline polyethylene, C$_{36}$H$_{74}$ (27°C) showing the intense lines in the low-frequency region (40 to 400 cm^{-1}) assigned to skeletal vibrations. (From Ref. 223.)

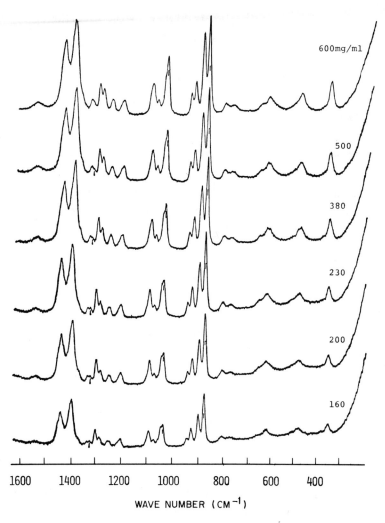

FIG. 51(b). Raman spectra of potassium n-butanoate in aqueous solutions at various concentrations (From Ref. 217.)

increase with n [215,216]. Moreover, the intensity of Raman scattering due to the accordion vibration increases with molecular concentration and is remarkably high at the critical micelle concentration.
This observation indicates that the percentage of all-trans conformers of the surfactant molecules is higher in the micellar aggregates
than in the case of isolated molecules [217].

The intensity of Raman scattering due to C-H stretching vibrations is also sensitive to changes in the environment of hydrocarbon
chains of lipid molecules. A correlation has been established [213,
214] between the extent of liquid-crystal formation and the ratio of
intensities of the Raman lines at 2850 and 2885 cm^{-1}, due respectively to symmetric stretching vibrations of the CH_2 and CH_3 groups
(Fig. 52). A similar correlation is observed for the intensity
ratio of the 2876 and 2937 cm^{-1} peaks in Raman spectra of potassium
n-hexane carbonates, where again the intensities change at the critical micelle concentration [218].

2. Phospholipids

The phospholipids occur widely in nature and are a major lipid
constituent of tissues. A typical phospholipid is phosphatidyl
choline (a lecithin), shown in structure XI, where R_1 and R_2 are
saturated or unsaturated fatty acid hydrocarbon chains.

$$
\begin{array}{c}
\quad\ \ \overset{\displaystyle O}{\overset{\displaystyle \|}{CH_2-O-C-R_1}} \\[2pt]
\overset{\displaystyle O}{\overset{\displaystyle \|}{R_2-C}}-O-CH \qquad\qquad\qquad CH_3 \\[2pt]
CH_2-O-\overset{O}{\underset{\underset{\ominus}{O}}{\overset{\|}{P}}}-O-CH_2CH_2-\overset{CH_3}{\underset{\overset{\oplus}{CH_3}}{N}}-CH_3
\end{array}
$$

XI

Other important phospholipids are phosphatidyl ethanolamine and
phosphatidyl serine, which differ from structure XI only in the
replacement of the choline residue $(-CH_2CH_2\overset{+}{N}(CH_3)_3)$ by ethanolamine
$(-CH_2CH_2\overset{+}{N}H_3)$ and serine $(-CH_2CH(\overset{+}{N}H_3)COO^-)$ residues, respectively.

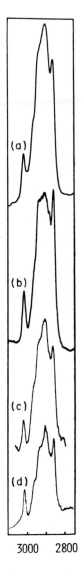

FIG. 52. Raman spectra showing the 2900-cm^{-1} region of the lamellar
(a) and cubic (b) liquid-crystalline phases of 1-monolinolein (10
and 25% by weight of water, respectively) and of lamellar (c) and
reversed hexagonal (d) liquid-crystalline phases of cardiolipin.
(From Ref. 214.)

a. *Infrared Spectra.* Vibrational spectra of phosphatidyl ethanolamine and related model compounds have been studied in detail [224]. Assignments are given in Table 16. Spectra of phospholipids generally exhibit major differences from spectra of the constituent fatty acids in regions where vibrations of the phosphodiester, glyceryl, and amino groups occur. The assignment of phosphate group vibrations is based upon studies of alkyl phosphates [5]. It will be noted that the OPO symmetric stretching frequency is assigned to the Raman line at 755 cm^{-1}, which is close to the value observed in spectra of dialkyl phosphate esters [5] but quite removed from the value of 814 cm^{-1} observed in polynucleotides and nucleic acids (Sec. III.B.2).

As shown in Fig. 53 and Table 16, the IR spectrum of phosphatidyl ethanolamine exhibits bands characteristic of the ionic groups, $\overset{+}{N}H_3$ and PO_2^-, indicating that both the solution and solid ("built-up film"; see Sec. 3,) samples contain the zwitterionic structure (XII) rather than the chelated structure (XIII) proposed by Abramson et al. [225].

$$\overset{\ominus}{O} \qquad \overset{\oplus}{NH_3}$$
$$-P-O-CH_2-CH_2$$
$$O$$

$$OH \cdots NH_2$$
$$-P-O-CH_2-CH_2$$
$$O$$

XII   XIII

Recent nmr evidence also favors structure XII [226].

IR spectra have also been used to detect a phase transition for the phosphatidyl ethanolamine-water system [227]. The transition occurs sharply in the interval 0 to 10°C, as detected by a plot of the IR intensity at 1470 cm^{-1} (CH_2 scissoring mode) vs. temperature, and presumably reflects the crystal to liquid-crystal phase transition. This phenomenon has been detected also in the phase transition of liposomes made of the phospholipids extracted from the membrane of temperature-sensitive mutants of *E. coli*. From plots of the intensity vs. temperature, the phase transition in liposomes of the mutant grown at 28°C is found to be 18°C. On the other hand for

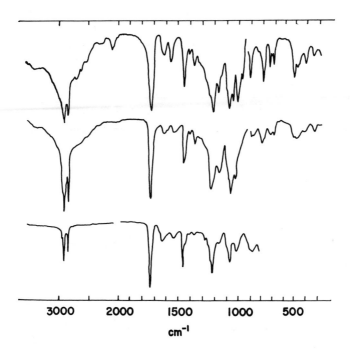

FIG. 53. IR spectra of phosphatidyl ethanolamine, solid (top), phosphatidyl ethanolamine in C_2Cl_4 (above 900 cm^{-1}, center) and in CS_2 solutions (below 900 cm^{-1}, center) and the built-up film of phosphatidyl ethanolamine (bottom). (From Ref. 224.)

mutants grown at 42°C, the corresponding phase transition occurs at 41°C. The difference of the transition temperature for the two kinds of mutants is mainly due to a difference in composition of their respective phospholipids [254].

   b. Raman Spectra. Raman spectra of dipalmitoyl lecithin [212] and of phosphatidyl ethanolamine [224] have been reported. The former is shown in Fig. 54(b) and contains lines due to the OPO (719 cm^{-1}) and PO_2^- (1090 cm^{-1}) groups, as well as more intense lines associated with the hydrocarbon chain [cf. Figs. 54(a) and (b)]. The position of the OPO frequency is considered indicative of the gauche-gauche conformation of the C-O-P-O-C network [5].

TABLE 16

Vibrational Spectra of Phosphatidylethanolamine[a]

IR (cm^{-1})	Raman (cm^{-1})		Assignments
750 (w)	755 (w)		O-P-O sym str
815 (w)			O-P-O antisym str
	875 (w)		C-C-N$^+$ sym str
890 (w)			CH$_2$ roc
980 (sh)			
1030 (m)			C-C-N$^+$ asym str
1068	1070 (sh)		C-O str, C-O-C sym str
	1075 (m)		and others
1090 (sh)	1097 (m)	(p)	PO$_2^-$ sym str
	1125 (w)		C-C str
1160 (m)			CH$_2$ wag, C-O-C antisym str and others
1228 (s)	1230 (w)	(dp)	PO$_2^-$ antisym str
	1236 (w)	(dp)	
	1310 (m)	(dp)	CH$_2$ wag, twist
1370 (w)			CH$_3$ sym def
1412 (w)			CH$_2$ sci, adjacent to C=O
	1440 (s)	(dp)	CH$_2$ sci
1460 (m)	1470 (sh)	(dp)	
1530 (w)			NH$_3^+$ sym def
1625 (w)			NH$_3^+$ deg def
1735 (s)			C=O str
2850 (s)	2855 (vs)	(p)	CH$_2$ sym str
2920 (vs)	2900 (s)		CH$_2$ antisym str
	2940 (vs)	(p)	NH$_3^+$ sym deg str, CH$_3$ sym str
2950 (s)			CH$_3$ deg str

[a]Abbreviations: s, strong; m, medium; w, weak, v, very; sh, shoulder; p, polarized; dp, depolarized; deg, degenerate; str, stretching; sym, symmetric; def, deformation; sci, scissoring; wag, wagging; roc, rocking; twist, twisting.

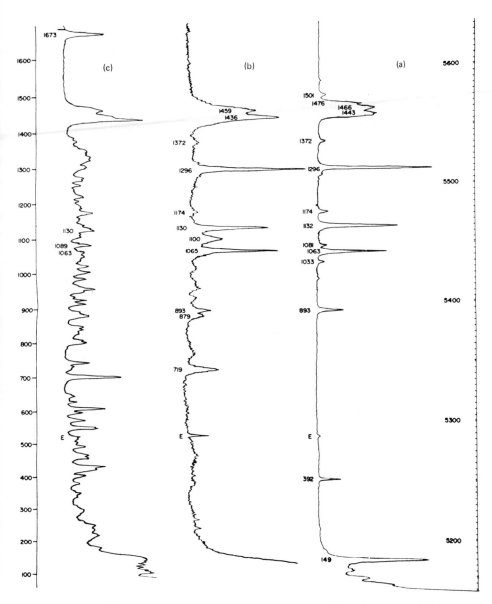

FIG. 54. Raman spectra of (a) polycrystalline hexadecane and (c) cholesterol, compared with the spectrum of (b) DL-dipalmitoyl lecithin monohydrate. (From Ref. 212.)

As stated above, fatty acids and phospholipids, which contain hydrophobic groups at one end of the molecule and hydrophilic groups at the other, tend to form micelles when in solution. In aqueous solutions, bi- or multilayer micelles are formed in which polar groups are exposed to the solvent. Transitions among different micelle structures or to an unstructured state may occur. Lippert and Peticolas [212] have measured the temperature dependence of certain Raman line intensities in spectra of aqueous lecithin multi-layers in an effort to observe such transitions. Fig. 55 shows for example the observed intensity change with temperature of the 1089-cm^{-1} line relative to that of the 1128-cm^{-1} line in spectra of di-palmitoyl lecithin. Similar results are reported for the intensity ratio 1089:1066 cm^{-1} as a function of temperature. For dipalmitoyl lecithin monohydrate and for lecithin multilayers, the gel liquid-crystal transition is found to be highly cooperative. However, addition of cholesterol in a 1:1 molar ratio changes the transition in the multilayers to a noncooperative one. It was suggested that a unique ordering of lecithin in the presence of water is required for the cooperative transition and that the absence of water or the addition of cholesterol disrupts this order [212]. Similar phenom-ena are observed for natural egg lecithin.

The results of Raman spectroscopy have been compared with the phase diagrams determined by physicochemical techniques, and it has been confirmed that the Raman lines clearly exhibit the change in state of the hydrocarbon chains [221,222]. The intensity ratio of the 1064-cm^{-1} and 1089-cm^{-1} peaks has also been used to detect the effects of various cations on the structure of phospholipids in micelles. Divalent cations apparently decrease the amount of gauche conformers in the order $Ba^{2+} < Mg^{2+} < Ca^{2+} \approx Cd^{2+}$ [220].

In addition to the Raman lines of the hydrocarbon chain near 1100 cm^{-1}, several other lines in this and other spectral regions are also found to be sensitive to phase transitions in phospholipid systems [219,221]. For example, frequency shifts occur in the PO_2^- symmetric stretching vibration near 1100 cm^{-1}; the intensity profile

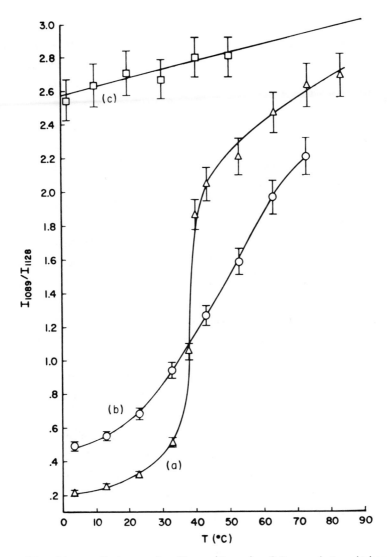

FIG. 55. Plots of the ratio (I_{1089}/I_{1128}) of Raman intensities at 1089 and 1128 cm^{-1} of DL-dipalmitoyl lecithin vs. temperature. (a) 20 wt % dipalmitoyl lecithin sonicated in H_2O; (b) 20 wt % dipalmitoyl lecithin-cholesterol (1:1 mixture) sonicated in H_2O; and (c) 10 wt % dipalmitoyl lecithin in CHCl$_3$. Similar results are obtained when I_{1089}/I_{1066}) is substituted for I_{1089}/I_{1128}. (From Ref. 212.)

in the CH stretching region is altered; and the accordion motion below 200 cm^{-1} is sensitive in position and intensity to such hydrocarbon chain melting. The last mentioned suggests that the hydrocarbon chains of lecithin are more folded than those of phosphatidylethanolamine.

3. *Multilayer Films*

When a drop of benzene or chloroform solution of a phospholipid is placed upon the surface of liquid water, the lipid molecules will spread to form a monolayer on the surface. If a plate of suitable IR window material is then dipped vertically into the liquid, the monolayer film will be adsorbed on the walls of the plate. Repitition of this procedure will ultimately produce a multilayer film of sufficient thickness to obtain an IR absorption spectrum (Fig. 56). A similar procedure may be used to obtain a multilayer film of a fatty acid.

Phospholipid multilayer films are often considered as model systems for the lipid phase in some cell membranes. Akutsu et al. [228] have obtained the IR spectrum shown in Fig. 53 for phosphatidylethanolamine deposited as a multilayer film on a plate of Irtran II. With use of IR radiation polarized in the z direction (Fig. 57), no dichroism is detected in either the x or y directions, indicating that the molecules in the multilayer film are oriented uniaxially. However, when the IR window is rotated by α_0 about the x axis, dichroism is detected for the x^1 and y^1 directions. The angle θ, through which the dipole transition moment M is tilted, is shown to be calculable from the equation

$$\frac{A^1_y}{A^1_x} = 1 + \frac{1}{n^2} \sin^2 \alpha_0 \, (2 \cos^2 \theta - 1) \tag{13}$$

where $A^1_{x_1}$ and $A^1_{y_1}$ are the absorbances of light polarized respectively in the x^1 and y^1 directions and n is the refractive index of the film [228]. It is thus found that the hydrocarbon chains are aligned at an angle of approximately 20° from the normal to the film surface.

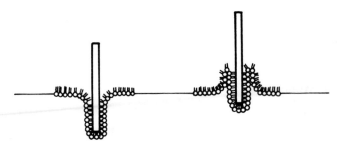

FIG. 56. Diagram showing the deposition of a phospholipid or fatty acid monolayer on the walls of a plate of IR window material. Repeated raising and lowering of the plate will ultimately result in the deposition of a multilayer film for transmission spectroscopy (see text).

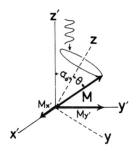

FIG. 57. Diagram showing the relative orientations of the electric vector of polarized IR radiation, the hydrocarbon chains in a phospholipid multilayer film, and the dipole transition moments. The angles α_o and θ are defined in the text, Eq. (13). (From Ref. 228.)

A model in which the phospholipid molecules are bent at the glyceryl residue with hydrocarbon chains and polar groups extending in mutually perpendicular directions is consistent with the IR data.

A phospholipid film suitable for IR transmission spectroscopy requires the deposition of about 100 monolayers on the IR window (Fig. 56). However, with reflection spectroscopy (ATR), 10 layers will usually suffice for a satisfactory spectrum. Müller et al. [229] have reported ATR spectra obtained by multiple internal reflections on 10-layer films of myristic acid. Comparison of the ATR data with surface tension measurements on myristic acid at

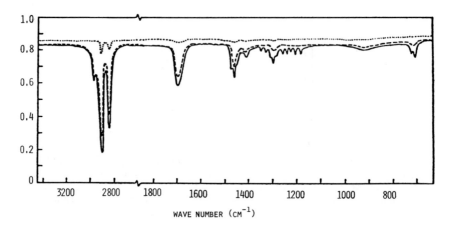

FIG. 58.   IR spectra of a built-up film of stearic acid (33 layers).
Solid curve, ATR spectrum with parallel polarization of the IR radiation; dashed curve, ATR spectrum with perpendicular polarization;
dotted curve, transmission spectrum.   (From Ref. 230.)

different pH values suggests that the hydrocarbon chains are in a
crystalline phase below pH 2.3 and in a fluid phase at higher pH.

ATR spectra of stearic acid films of different thicknesses
have been obtained by Takenaka et al. [230].   Band progressions of
the CH_2 wagging and twisting vibrations are clearly evident between
1150 and 1350 cm^{-1}, as shown in Fig. 58.   The absence of a band in
the OH stretching region suggests that carboxylic acid dimers may
be formed with consequent lowering of the OH stretching frequency
and broadening of the expected band.

4. Steroids

Steroids are lipids which contain the cyclopentanoperhydro-
phenanthrene nucleus (XIV).   Among them are such important materials
as the bile acids, androgens, estrogens, adrenocortical hormones,
and cholesterol (XV).

The steroids have been extensively studied by IR spectroscopy
which has been used primarily to identify substituents in the nucleus

XIV   XV

and to locate sites of unsaturation. Several reviews of the considerable literature on IR applications to steroid research are available
[1,231-235].

An important steroid in the biochemistry of man is cholesterol
(XV), which is thought to play a role in the stability and rigidity
of lipid phases in human tissues. As mentioned earlier (Sec. 2.b),
evidence from Raman spectra [212,222] suggests that cholesterol may
decrease the interactions between paraffin side chains of the phospholipid, dipalmitoyl lecithin.

IR spectroscopy of cholesterol has been directed mainly at detecting its hydrogen-bonding interactions with other lipids. Zull
et al. [236], for example, obtained the ATR spectrum of the film
cast from a CCl_4 solution of cholesterol-lecithin (1:1 molar mixture)
and compared it with a summation spectrum (Fig. 59). The OH stretching frequency of cholesterol at 3400 cm^{-1} shifts to 3250 cm^{-1} in the
mixture. Although the reason for the shift was not clearly stated,
it was suggested that the phosphate oxygens of lecithin may be involved in hydrogen bonding to cholesterol donor groups.

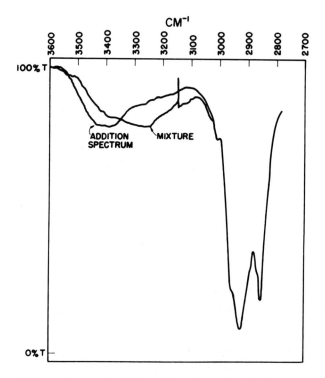

FIG. 59.   ATR spectrum of a 1:1 molar mixture of cholesterol and
lecithin compared with a summation spectrum of lecithin and cholesterol.   (From Ref. 236.)

    Parker and Bhaskar [237] have given a quantitative study of the
hydrogen bonding between cholesterol and various triglycerides in
CCl_4 solution.   The OH group of cholesterol is hydrogen bonded to
the ester carbonyl oxygen in triglycerides though the bonding is not
apparently as strong as that between the cholesterol molecules themselves.

    5.   *Membranes*

    Lipids and proteins are the major constituents of biological
cell membranes.   IR spectra of membranes thus resemble composite
spectra of these constituents.   A typical membrane spectrum is shown
in Fig. 60.

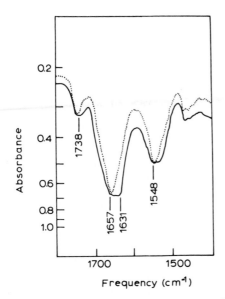

FIG. 60.   IR spectra of solid films of outer (solid curve) and cytoplasmic (broken curve) membranes.   (From Ref. 248.)

Most IR studies of membranes have been concerned with determining the conformations of the protein components from the position of
the amide I band. As is evident from Fig. 60 and the foregoing discussions, the amide I band is the most intense of the observable
peaks in the IR spectrum of a membrane and is relatively well isolated from absorption bands due to other functional groups of protein
and lipid components. In a variety of membranes including erythrocyte ghosts [238,239], cytoplasmic and endoplasmic reticula of
Ehrlich ascites cells [240], myelin [241], *Micrococcus lysodeikticus*
membrane [247], and cytoplasmic membrane of *Escherichia coli* [248]
the amide I band is centered near 1655 cm^{-1} with no distinguishable
band or shoulder near 1630 cm^{-1}. These results are not changed by
extraction of the lipid components. It is thus concluded that the
proteins of these membranes do not occur in the β conformation but
are in either α-helical or random chain conformations. These findings
cast doubt on proposed models for membranes in which the proteins
were considered to form an extended sheet structure (β type).

In contrast, the IR spectra of membranes from rat liver [243], microplasma [244], rat adipocyte plasma [245], and outer membrane of *Escherichia coli* [248] reveal amide I absorption near 1630 cm^{-1} suggesting the presence of appreciable β structure in the protein components. Reconstituted membrane of *Micrococcus lysodeikticus* [242] and the recombinants of erythrocyte apoprotein and lipid bilayers [246] also display an amide I feature at 1630 cm^{-1}. It has also been claimed that various chemical treatments of plasma membrane (e.g., exposure to acidic media, ATP hydrolysis, etc. [247]) can bring about a change in protein conformation of the β structure. However, such structural conclusions from the spectra should be reached only after other possible causes of the 1630-cm^{-1} shoulder (such as uncompensated water vapor bands, intermolecular aggregations, etc.) can be definitively excluded.

IR spectra have also been used to investigate the interactions of lipid components in membranes. For example, Jenkinson et al. [241] observed a band at 720 cm^{-1} in IR spectra of myelin and its lipid extracts and assigned it to the CH_2 rocking vibration of the all-trans hydrocarbon chain.

Raman spectra have also been obtained on heme-free erythrocyte ghosts [249-252]. The conformation of protein within the membrane, as estimated from the observed Raman frequencies in the amide III region (1240 to 1270 cm^{-1}), contains 40 to 55% α helix. The hydrocarbon side chains of the phospholipid component, as interpreted from the 1060- to 1130-cm^{-1} C-C stretching region, are estimated to contain 60% of the all-trans configuration [252]. In these specimens fairly strong Raman lines are sometimes observed at 1530 and 1165 cm^{-1}, presumably due to membrane-bound β-carotenoids. The intensities of these bands depends on various experimental factors associated with the state of the membrane [250,251].

Fluorescent impurities make it difficult to obtain Raman spectra of satisfactory quality on membranes. However, to the extent that this difficulty can be circumvented in the years ahead, it is anticipated that laser Raman spectra will be able to provide more interesting information on the structure and function of membrane systems.

Acknowledgments. This review was initially prepared by the authors
at their respective institutions in 1973. It was revised and up-
dated in 1975 when the first author was a visiting scientist at the
Institute for Protein Research, Osaka University. The support of
the National Science Foundation (Grant OIP 74-12149, under the
United States-Japan Cooperative Science Program) and the Southeastern
Massachusetts University (Sabbatical Leave), which made possible the
visit to Japan, are gratefully acknowledged. We also thank the
Ministry of Education (Japan) and the Research Corporation (United
States) for support and Mrs. M. Roberts (S.M.U.) and Miss F. Osaki
(Osaka University) for assistance in typing the manuscript.

REFERENCES

1. F. S. Parker, *Applications of Infrared Spectroscopy in Biochem-
    istry, Biology and Medicine,* Plenum Press, New York, 1971.

2. G. J. Thomas, Jr., in *Physical Techniques in Biological Research,*
    vol. 1A, 2nd ed. (G. Oster, ed.), Academic Press, New York, 1971.

3. H. Susi, in *Structure and Stability of Biological Macromolecules,*
    vol. III (S. N. Timasheff and G. Fasman, eds.), Marcel Dekker,
    New York, 1969.

4. J. L. Koenig, *J. Polym. Sci,* Part D, *1972,* 59.

5. B. G. Frushour and J. L. Koenig, "Raman Spectroscopy of Proteins,"
    in *Advances in Infrared and Raman Spectroscopy* (R. J. H. Clark
    and R. E. Hester, eds.), Heyden & Son, Ltd., London, 1975.

6. T. Shimanouchi, M. Tsuboi, and Y. Kyogoku, Adv. Chem. Phys., *7,*
    435 (1964).

7. M. Tsuboi, *Appl. Spectrosc., 3,* 45 (1969).

8. M. Tsuboi and Y. Kyogoku, in *Synthetic Procedures in Nucleic
    Acid Chemistry,* Vol. II (W. W. Zorbach and S. Tipson, eds.),
    Wiley, New York, 1973.

9. M. Tsuboi, in *Basic Principles in Nucleic Acid Chemistry,* Vol. I
    (P. O. P. Ts'o, ed.), Academic Press, New York, 1973.

10. M. Tsuboi, S. Takahashi, and I. Harada, in *Physico-Chemical
    Properties of Nucleic Acids* (J. Duchesne, ed.), Academic Press,
    New York, 1973.

11. K. A. Hartman, R. C. Lord, and G. J. Thomas, Jr., in *Physico-
    Chemical Properties of Nucleic Acids* (J. Duchesne, ed.), Aca-
    demic Press, New York, 1973.

12. G. J. Thomas, Jr., in *Vibrational Spectra and Structure,* vol. 3
    (J. Durig, ed.), Marcel Dekker, New York, 1974.

13. T. G. Spiro, in *Chemical and Biochemical Applications of Lasers*, (C. B. Moore, ed.), Academic Press, New York, 1974.

14. H. J. Bernstein, *Resonance Raman Spectroscopy: A Review* (in press, 1977).

15. W. J. Potts, Jr., *Chemical Infrared Spectroscopy, Vol. I: Techniques.*, Wiley, New York, 1963.

16. R. J. Bell, *Introductory Fourier Transform Spectroscopy*, Academic Press, New York, 1972.

17. J. O. Alben, G. H. Bare, and P. A. Bromberg, *Nature, 252,* 736 (1974).

18. G. H. Bare, J. O. Alben, and P. A. Bromberg, *Biochemistry, 14,* 1578 (1975).

19. M. Falk, K. A. Hartman, and R. C. Lord, *J. Amer. Chem. Soc., 84,* 3843 (1962).

20. G. B. B. M. Sutherland and M. Tsuboi, *Proc. Roy. Soc., 239,* 446 (1957).

21. E. R. Blout and H. Lenormant, *J. Opt. Soc. Amer., 43,* 1093 (1953).

22. H. T. Miles, *Methods Enzymol., 12,* Part B, 256 (1968).

23. R. C. Lord and T. J. Porro, *Zeits. Elektrochem., 64,* 672 (1960).

24. M. Falk, K. A. Hartman, and R. C. Lord, *J. Amer. Chem. Soc., 85,* 387 (1963).

25. N. J. Harrick, *Internal Reflection Spectroscopy*, Wiley-Interscience, New York, 1967.

26. H. A. Szymanski, ed., *Raman Spectroscopy*, Plenum Press, New York, Vol. 1: 1967; vol. 2: 1970.

27. E. Loader, *Basic Laser Raman Spectroscopy*, Heyden & Son, Ltd., London, 1970.

28. M. C. Tobin, *Laser Raman Spectroscopy*, Wiley-Interscience, New York, 1971.

29. J. A. Koningstein, *Introduction to the Theory of the Raman Effect*, D. Reidel Publishing Co., Dordrecht-Holland, 1972.

30. S. K. Freeman, *Applications of Laser Raman Spectroscopy.* Wiley-Interscience, New York, 1974.

31. A. Anderson, ed., *The Raman Effect*, Marcel Dekker, New York, Vol. 1: 1971; Vol. 2: 1973.

32. G. Careri, V. Mazzacurati, and G. Signorelli, *Phys. Lett., 31A,* 425 (1970).

33. G. J. Thomas, Jr., G. C. Medeiros, and K. A. Hartman, *Biochem. Biophys. Res. Commun., 44,* 587 (1971).

34. G. J. Thomas, Jr. and J. R. Barylski, *Appl. Spectrosc., 24,* 463 (1970).

35. E. W. Small and W. L. Peticolas, *Biopolymers, 10,* 69 (1971).

36. J. W. Amy, R. W. Chrisman, J. W. Lundeen, T. Y. Ridley, J. C. Sprowles, and R. S. Tobias, *Appl. Spectrosc., 28,* 262 (1974).

37. L. Nafie, P. Stein, B. Fanconi, and W. L. Peticolas, *J. Chem. Phys., 52,* 1584 (1970).

38. W. T. Wilser, Ph.D. thesis, Materials Science Center, Cornell University, Ithaca, New York, 1975 (D. B. Fitchen, Thesis Director).

39. W. T. Wilser and D. B. Fitchen, *Biopolymers, 13,* 1435 (1974).

40. W. T. Wilser and D. B. Fitchen, *J. Chem. Phys., 62,* 720 (1975).

41. H. J. Bernstein and G. Allen, *J. Opt. Soc. Amer., 45,* 237 (1955).

42. D. G. Rea, *J. Mol. Spectrosc., 4,* 507 (1960).

43. D. D. Tunnicliff and A. C. Jones, *Spectrochim. Acta, 18,* 579 (1962).

44. R. N. Jones, J. B. DiGiorgio, J. J. Elliott, and G. A. A. Nonnenmacher, *J. Org. Chem., 30,* 1822 (1965).

45. S. K. Freeman and D. O. Landon, *Anal. Chem., 41,* 398 (1969).

46. R. E. Hester, *Anal. Chem., 44,* 490R (1972).

47. T. Miyazawa, in *Polyamino acids, Polypeptides and Proteins* (M. A. Stahmann, ed.), Univ. Wisconsin Press, Madison, 1962.

48. J. Jakes and Krimm, *Spectrochim. Acta, 27A,* 19, 35 (1971).

49. Y. Abe and S. Krimm, *Biopolymers, 11,* 1817, 1841 (1972).

50. S. Krimm and Y. Abe, *Proc. Nat. Acad. Sci. U.S.A., 69,* 2788 (1972).

51. T. Miyazawa, *J. Chem. Phys., 32,* 1647 (1960).

52. T. Miyazawa and E. R. Blout, *J. Amer. Chem. Soc., 83,* 712 (1961).

53. S. Krimm, *J. Mol. Biol., 4,* 528 (1962).

54. Y. N. Chirgadze and E. V. Brazhnikov, *Biopolymers, 13,* 1701 (1974).

55. Y. N. Chirgadze, O. V. Fedorov, and N. P. Trushina, *Biopolymers, 14,* 679 (1975).

56. H. Susi, S. N. Timasheff, and L. Stevens, *J. Biol. Chem., 242,* 5460 (1967).

57. B. Fanconi, B. Tomlinson, L. A. Nafie, E. W. Small, and W. L. Peticolas, *J. Chem. Phys., 51,* 3993 (1969).

58. R. C. Lord, *Proc. Int. Congr. Pure Appl. Chem. Suppl. 23rd, 7,* 179 (1971).

59. R. C. Lord and N. T. Yu, *J. Mol. Biol., 50,* 509 (1970).

60. L. Simons, G. Bergstrom, G. Blomfelt, S. Forss, H. Stenback, and G. Wansen, *Comm. Phys.-Math., 42,* 125 (1972).

61. H. Sugeta, A. Go, and T. Miyazawa, *Chem. Lett.* (Japan), *1972*, 83; *Bull. Chem. Soc. Japan, 46,* 3407 (1973).

62. H. Sugeta, *Spectrochim. Acta, 31A,* 1729 (1975).

63. N. Nogami, H. Sugeta, and T. Miyazawa, *Chem. Lett., 1975,* 147.

64. Y. Ozaki, H. Sugeta, and T. Miyazawa, *Chem. Lett., 1975,* 713.

65. H. E. Van Wart, A. Lewis, H. A. Scheraga, and F. D. Saeva, *Proc. Natl. Acad. Sci. U.S., 70,* 2619 (1973).

66. M. Nakanishi, H. Takesada, and M. Tsuboi, *J. Mol. Biol., 89,* 241 (1974).

67. P. Sutton and J. L. Koenig, *Biopolymers, 9,* 615 (1970).

68. M. Smith, A. G. Walton, and J. L. Koenig, *Biopolymers, 8,* 29 (1969).

69. J. L. Koenig and B. Frushour, *Biopolymers, 11,* 1871 (1972).

70. K. Itoh, T. Hinomoto, and T. Shimanouchi, *Biopolymers, 13,* 307 (1974).

71. M. C. Chen and R. C. Lord, *J. Amer. Chem. Soc., 96,* 4750 (1974).

72. T-J. Yu, J. L. Lippert, and W. L. Peticolas, *Biopolymers, 12,* 2161 (1973).

73. J. L. Koenig and P. L. Sutton, *Biopolymers, 8,* 167 (1969).

74. R. C. Lord and N. T. Yu, *J. Mol. Biol., 51,* 203 (1970).

75. A. M. Bellocq, R. C. Lord, and R. Mendelsohn, *Biochim. Biophys. Acta, 257,* 280 (1972).

76. R. C. Lord and R. Mendelsohn, *J. Amer. Chem. Soc., 94,* 2133 (1970).

77. M. C. Chen, R. C. Lord, and R. Mendelsohn, *J. Amer. Chem. Soc., 96,* 3038 (1974).

78. N. T. Yu and B. H. Jo, *Arch. Biochem. Biophys., 156,* 469 (1973).

79. N. T. Yu, C. S. Liu, J. Culver, and D. C. O'Shea, *Biochim. Biophys. Acta, 263,* 1 (1972).

80. N. T. Yu, C. S. Liu, and D. C. O'Shea, *J. Mol. Biol., 70,* 117 (1972).

81. M. C. Chen, R. C. Lord, and R. Mendelsohn, *Biochim. Biophys. Acta, 328,* 252 (1973).

82. N. T. Yu and C. S. Liu, *J. Amer. Chem. Soc., 94,* 3250, 5127 (1972).

83. N. T. Yu, B. H. Jo, and C. S. Liu, *J. Amer. Chem. Soc., 94,* 7572 (1972).

84. N. T. Yu and B. H. Jo., *J. Amer. Chem. Soc., 95,* 5033 (1973).

85. N. T. Yu, B. H. Jo, and D. C. O'Shea, *Arch. Biochem. Biophys., 156,* 71 (1973).

86. N. T. Yu, T-S. Lin, and A. T. Tu, *J. Biol. Chem.*, *250*, 1782 (1975).

87. G. J. Thomas, Jr., B. Prescott, P. McDonald-Ordzie, and K. A. Hartman, *J. Mol. Biol.*, *102*, 103 (1976).

88. G. J. Thomas, Jr. and P. Murphy, *Science*, *188*, 1205 (1975).

89. J. M. Eyster and E. W. Prohofsky, *Biopolymers*, *13*, 2505 (1974).

90. T. V. Long, II, T. M. Loehr, J. R. Allkins, and W. Lovenberg, *J. Amer. Chem. Soc.*, *93*, 1809 (1971).

91. A. Lewis, J. Spoonhower, R. A. Bogomolni, R. H. Lozier, and W. Stoeckenius, *Proc. Natl. Acad. Sci. U.S.A.*, *71*, 4462 (1974).

92. V. Miskowski, S.-P. W. Tang, T. G. Spiro, E. Shapiro; and T. H. Moss, *Biochemistry*, *14*, 1244 (1975).

93. M. Lutz, C. R., *Acad. Sci. Paris*, *275B*, 497 (1972).

94. M. Lutz, *J. Raman Spectrosc.*, *2*, 497 (1974).

95. S. P. W. Tang, T. G. Spiro, K. Mukai, and T. Kimura, *Biochem. Biophys. Res. Commun.*, *53*, 869 (1973).

96. J. B. R. Dunn, D. F. Shriver, and I. M. Klotz, *Biochemistry*, *14*, 2689 (1975).

97. T. C. Strekas and T. G. Spiro, *Biochim. Biophys. Acta*, *263*, 830 (1972).

98. H. Brunner, A. Mayer, and H. Sussner, *J. Mol. Biol.*, *70*, 153 (1972).

99. H. Brunner and H. Sussner, *Biochim. Biophys. Acta*, *310*, 20 (1973).

100. T. G. Spiro and T. C. Strekas, *Proc. Natl. Acad. Sci. U.S.A.*, *69*, 2622 (1972).

101. T. G. Spiro and T. C. Strekas, *J. Amer. Chem. Soc.*, *96*, 338 (1974).

102. T. C. Strekas and T. G. Spiro, *J. Raman Spectrosc.*, *1*, 387 (1973).

103. W. H. Woodruff, T. G. Spiro, and T. Yonetani, *Proc. Natl. Acad. Sci. U.S.A.*, *71*, 1065 (1974).

104. T. C. Strekas and T. G. Spiro, *Biochim. Biophys. Acta*, *278*, 188 (1972).

105. T. G. Spiro, *Biochim. Biophys. Acta*, *416*, 169 (1975).

106. T. Yamamoto, G. Palmer, D. Gill, I. Salmeen, and L. Rimai, *J. Biol. Chem.*, *248*, 5211 (1973).

107. I. Salmeen, L. Rimai, D. Gill, T. Yamamoto, G. Palmer, C. R. Hartzell, and H. Beinert, *Biochem. Biophys. Res. Commun.*, *52*, 1100 (1973).

108. T. M. Loehr and J. S. Loehr, *Biochem. Biophys. Res. Commun.*, *55*, 218 (1973).

109. D. W. Collins, D. B. Fitchen, and A. Lewis, *J. Chem. Phys.*, *59*, 5714 (1973).

110. M. Pezolet, L. A. Nafie, and W. L. Peticolas, *J. Raman Spectrosc.*, *1*, 455 (1973).

111. J. Nestor and T. G. Spiro, *J. Raman Spectrosc.*, *1*, 539 (1973).

112. T. C. Strekas and T. G. Spiro, *Biochim. Biophys. Acta*, *351*, 237 (1974).

113. M. Ikeda-Saito, T. Kitagawa, T. Iizuka, and Y. Kyogoku, *FEBS Lett.*, *50*, 233 (1975).

114. T. Kitagawa, Y. Kyogoku, T. Iizuka, M. Ikeda-Saito, and T. Yamanaka, *J. Biochem.*, *78*, 719 (1975).

115. T. Kitagawa, T. Iizuka, M. Saito, and Y. Kyogoku, *Chem. Lett.*, *1975*, 849.

116. Y. Ozaki, T. Kitagawa, and Y. Kyogoku, *FEBS Lett.*, *62*, 369 (1976).

117. T. Kitagawa, Y. Orii, and Y. Kyogoku, *Arch. Biochem. Biophys.*, *180*, 2 (1977).

118. Y. Ozaki, T. Kitagawa, Y. Kyogoku, H. Ogoshi, E. Watanabe, and Z. Yoshida, unpublished.

119. A. L. Verma and H. J. Bernstein, *Biochem. Biophys. Res. Commun.*, *57*, 255 (1974).

120. A. L. Verma, R. Mendelsohn, and H. J. Bernstein, *J. Chem. Phys.*, *61*, 383 (1974).

121. A. L. Verma and H. J. Bernstein, *J. Raman Spectrosc.*, *2*, 163 (1974).

122. R. H. Felton, N. T. Yu, D. C. O'Shea, and J. A. Shelnutt, *J. Amer. Chem. Soc.*, *96*, 3675 (1974).

123. L. D. Spaulding, C. C. Chang, N. T. Yu, and R. H. Felton, *J. Amer. Chem. Soc.*, *97*, 2517 (1975).

124. T. Kitagawa, H. Ogoshi, E. Watanabe, and Z. Yoshida, *Chem. Phys. Lett.*, *30*, 451 (1975).

125. T. Kitagawa, H. Ogoshi, E. Watanabe, and Z. Yoshida, *J. Phys. Chem.*, *79*, 2629 (1976).

126. T. Kitagawa, M. Abe, Y. Kyogoku, H. Ogoshi, E. Watanabe, and Z. Yoshida, *J. Phys. Chem.*, *80*, 1181 (1976).

127. P. Stein, J. M. Burke, and T. G. Spiro, *J. Amer. Chem. Soc.*, *97*, 2304 (1975).

128. M. Abe, T. Kitagawa, and Y. Kyogoku, *Chem. Lett.*, 249 (1976). lished.

129. H. Ogoshi, Y. Saito, and K. Nakamoto, *J. Chem. Phys.*, *57*, 4194 (1972).

130. P. R. Carey, H. Schneider, and H. J. Bernstein, *Biochem. Biophys. Res. Commun.*, *47*, 588 (1972).

131. P. R. Carey, A. Froese, and H. Schneider, *Biochemistry, 12*, 2198 (1973).

132. H. T. Miles, *Proc. Nat. Acad. Sci.*, *U.S.A.*, *47*, 791 (1971).

133. R. V. Wolfenden, *J. Mol. Biol.*, *40*, 307 (1969).

134. G. C. Medeiros and G. J. Thomas, Jr., *Biochim. Biophys. Acta*, *238*, 1 (1971).

135. R. J. Steiner and R. J. Beers, *Polynucleotides*, Elsevier, New York, 1961, Chap. 2.

136. R. C. Lord and G. J. Thomas, Jr., *Spectrochim. Acta, 23A*, 2551 (1967).

137. M. Tomasz, J. Olson, and C. M. Mercado, *Biochemistry, 11*, 1235 (1972).

138. A. Psoda and D. Shugar, *Biochim. Biophys. Acta*, *247*, 507 (1971).

139. G. C. Medeiros and G. J. Thomas, Jr., *Biochim. Biophys. Acta*, *247*, 449 (1971).

140. R. C. Lord and G. J. Thomas, Jr., *Dev. Appl. Spectrosc.*, *6*, 179 (1968).

141. G. J. Thomas, Jr. and J. Livramento, *Biochemistry, 14*, 5210 (1975).

142. R. L. Sinsheimer, R. L. Nutter, and G. R. Hopkins, *Biochim. Biophys. Acta, 18*, 13 (1955).

143. L. Rimai, T. Cole, J. L. Parsons, J. T. Hickmott, Jr., and E. B. Carew, *Biophys. J.*, *9*, 320 (1969).

144. L. Rimai and M. E. Heyde, *Biochem. Biophys. Res. Commun.*, *41*, 313 (1970).

145. R. M. Hamlin, Jr., R. C. Lord, and A. Rich, *Science, 148*, 1734 (1965).

146. Y. Kyogoku, R. C. Lord, and A. Rich, *J. Amer. Chem. Soc.*, *89*, 496 (1967).

147. Y. Kyogoku, R. C. Lord, and A. Rich, *Proc. Nat. Acad. Sci. U.S.A.*, *57*, 250 (1967).

148. Y. Kyogoku, R. C. Lord, and A. Rich, *Biochim. Biophys. Acta*, *179*, 10 (1969).

149. Y. Kyogoku, B. S. Yu, R. C. Lord, and A. Rich, in *The Purines -- Theory and Experiment* (Fourth Jerusalem Symposium on Quantum Chemistry and Biochemistry), The Israel Academy of Sciences and Humanities, Jerusalem, 1972, p. 223.

150. Y. Kyogoku, R. C. Lord, and A. Rich, *Nature, 218*, 69 (1968).

151. N. T. Yu and Y. Kyogoku, *Biochim. Biophys. Acta, 331*, 21 (1973).

152. Y. Kyogoku and B. S. Yu, *Bull. Chem. Soc. Japan*, *41*, 1742 (1968).

153. Y. Kyogoku and B. S. Yu, *Bull. Chem. Soc. Japan*, *42*, 1387 (1969).

154. Y. Kyogoku and B. S. Yu, *Chem.-Biol. Interactions*, *2*, 117 (1970).

155. B. S. Yu and Y. Kyogoku, *Bull. Chem. Soc. Japan*, *43*, 239 (1970).

156. R. C. Lord and G. J. Thomas, Jr., *Biochim. Biophys. Acta*, *142*, 1 (1967).

157. M. P. Schweizer, A. D. Broom, P. O. P. Ts'o, and D. P. Hollis, *J. Amer. Chem. Soc.*, *90*, 1043 (1968).

158. M. Tsuboi, *Proc. Int. Congr. Pure Appl. Chem. Suppl. 23rd*, *7*, 145 (1971).

159. E. W. Small and W. L. Peticolas, *Biopolymers*, *10*, 1377 (1971).

160. J. Rice, L. Lafleur, G. C. Medeiros, and G. J. Thomas, Jr., *J. Raman Spectrosc.*, *1*, 207 (1973).

161. B. Prescott, R. Gamache, J. Livramento, and G. J. Thomas, Jr., *Biopolymers*, *13*, 1821 (1974).

162. K. A. Hartman, *Biochim. Biophys. Acta*, *138*, 192 (1967).

163. S. Mansy, T. E. Wood, J. C. Sprowles, and R. S. Tobias, *J. Amer. Chem. Soc.*, *96*, 1762 (1974).

164. S. Mansy and R. S. Tobias, *J. Amer. Chem. Soc.*, *96*, 6874 (1974).

165. A. T. Tu and M. J. Heller, in *Metal Ions in Biological Systems*, Vol. I (H. Sigel, ed.), Marcel Dekker, New York, 1974.

166. M. C. Chen, R. Geige, R. C. Lord, and A. Rich, *Biochemistry*, *14*, 4385 (1975).

167. M. Heyde and L. Rimai, *Biochemistry*, *10*, 1121 (1971).

168. Y. Kyogoku, S. Higuchi, and M. Tsuboi, *Spectrochim. Acta*, *23A*, 969 (1969).

169. I. Harada and R. C. Lord, *Spectrochim. Acta*, *26A*, 2305 (1970).

170. M. Tsuboi, *J. Polym. Sci.*, Part C, *7*, 125 (1964).

171. G. J. Thomas, Jr., *Biochim. Biophys. Acta*, *213*, 417 (1969).

172. G. J. Thomas, Jr. and K. A. Hartman, *Biochim. Biophys. Acta*, *312*, 311 (1973).

173. F. Ishikawa, J. Frazier, F. B. Howard, and H. T. Miles, *J. Mol. Biol.*, *70*, 475 (1972).

174. H. T. Miles and J. Frazier, *Biochem. Biophys. Res. Commun.*, *14*, 21 (1964).

175. L. Lafleur, J. Rice, and G. J. Thomas, Jr., *Biopolymers*, *11*, 2423 (1972).

176. G. J. Thomas, Jr., *Biopolymers, 7*, 325 (1969).

177. K. Morikawa, M. Tsuboi, S. Takahashi, Y. Kyogoku, Y. Mitsui, Y. Iitaka, and G. J. Thomas, Jr., *Biopolymers, 12*, 799 (1973).

178. N. T. Yu, Ph.D. thesis, submitted to the Department of Chemistry, M.I.T., Cambridge, Mass., 1968.

179. K. A. Hartman and G. J. Thomas, Jr., *Science, 170*, 740 (1970).

180. K. G. Brown, E. J. Kiser, and W. L. Peticolas, *Biopolymers, 11*, 1855 (1972).

181. G. J. Thomas, Jr., G. C. Medeiros, and K. A. Hartman, *Biochim. Biophys. Acta, 277*, 71 (1972).

182. G. J. Thomas, Jr., M. C. Chen, and K. A. Hartman, *Biochim. Biophys. Acta, 324*, 37 (1973).

183. K. A. Hartman, N. Clayton, and G. J. Thomas, Jr., *Biochem. Biophys. Res. Commun., 50*, 942 (1973).

184. S. C. Erfurth, E. J. Kiser, and W. L. Peticolas, *Proc. Natl. Acad. Sci. U.S.A., 69*, 938 (1972).

185. F. M. Pohl, A. Ranade, and M. Stackburger, *Biochim. Biophys. Acta, 335*, 85 (1975).

186. G. J. Thomas, Jr., presented at the Symposium on the Spectrum of Light Scattered from Biological Molecules, M.I.T., Cambridge, Mass., April 10-11, 1972.

187. Y. Nishimura, K. Morikawa, and M. Tsuboi, *Bull. Chem. Soc. Japan, 47*, 1043 (1974).

188. E. W. Small, K. G. Brown, and W. L. Peticolas, *Biopolymers, 11*, 1209 (1972).

189. M. C. Chen and G. J. Thomas, Jr., *Biopolymers, 13*, 615 (1974).

190. G. J. Thomas, Jr., M. C. Chen, R. C. Lord, P. S. Kotsiopoulos, T. R. Tritton, and S. C. Mohr, *Biochem. Biophys. Res. Commun., 54*, 570 (1973).

191. G. J. Thomas, Jr. and M. Spencer, *Biochim. Biophys. Acta, 179*, 360 (1969).

192. M. C. Tobin, *Spectrochim. Acta, 25*, 1855 (1969).

193. S. Arnott, *Prog. Biophys. Mol. Biol., 21*, 265 (1970).

194. L. Rimai, V. M. Maher, D. Gill, I. Salmeen, and J. J. McCormick, *Biochim. Biophys. Acta, 361*, 155 (1974).

195. S. C. Erfurth and W. L. Peticolas, *Biopolymers, 14*, 247 (1975).

196. F. J. Bullock and O. Jardetzky, *J. Org. Chem., 29*, 1988 (1964).

197. M. P. Schweizer, S. I. Chan, G. K. Helmkamp, and P. O. P. Ts'o, *J. Amer. Chem. Soc., 86*, 696 (1964).

198. R. N. Maslova, E. A. Lesnik, and Ya. M. Varshavsky, *Molekulyarnaya Biologiya, 3*, 728 (1969); Biochem. Biophys. Res. Commun., *34*, 260 (1969).

199. R. C. Gamble and P. R. Schimmel, *Proc. Natl. Acad. Sci. U.S.A.*, *71*, 1356 (1974).

200. J. Livramento and G. J. Thomas, Jr., *J. Amer. Chem. Soc.*, *96*, 6529 (1974).

201. R. N. Maslova, E. A. Lesnik, and Ya. M. Varshavsky, *Biochem. Biophys. Res. Commun.*, *34*, 260 (1969).

202. M. Tsuboi, S. Takahashi, S. Muraishi, and T. Kajiura, *Bull. Chem. Soc. Japan*, *44*, 2921 (1971).

203. M. Tsuboi, A. Y. Hirakawa, Y. Nishimura, and J. Harada, *J. Raman Spectrosc.*, *2*, 609 (1974).

204. A. Y. Hirakawa and M. Tsuboi, *Science*, *188*, 359 (1975).

205. M. Pézolet, T. J. Yu, and W. L. Peticolas, *J. Raman Spectrosc.*, *3*, 55 (1975).

206. R. I. Cotter and W. B. Gratzer, *Eur. J. Biochem.*, *8*, 352 (1969).

207. C. A. Knight, *Chemistry of Viruses*, 2nd ed., Springer-Verlag, 1975.

208. R. G. Sinclair, A. F. McKay, and R. N. Jones, *J. Amer. Chem. Soc.*, *74*, 2570 (1952).

209. J. L. Lippert and W. L. Peticolas, *Biochim. Biophys. Acta*, *282*, 8 (1972).

210. R. G. Snyder, *J. Mol. Spectrosc.*, *4*, 411 (1960).

211. R. F. O'Connor, *J. Amer. Oil. Chem. Soc.*, *23*, 1 (1956).

212. J. L. Lippert and W. L. Peticolas, *Proc. Natl. Acad. Sci. U.S.A.*, *68*, 1572 (1971).

213. K. Larsson, *Chem. Phys. Lipids*, *10*, 165 (1973).

214. K. Larsson and R. P. Rand, *Biochim. Biophys. Acta*, *326*, 245 (1973).

215. D. Chapman, *J. Chem. Soc.*, *1957*, 4489.

216. H. Okabayashi, M. Okuyama, T. Kitagawa, and T. Miyazawa, *Bull. Chem. Soc. Japan*, *47*, 1075 (1974).

217. H. Okabayashi, M. Okuyama, and T. Kitagawa, *Bull. Chem. Soc. Japan*, *48*, 2264 (1975).

218. H. Okabayashi and T. Kitagawa, unpublished.

219. K. G. Brown, W. L. Peticolas, and E. Brown, *Biochim. Biophys. Res. Commun.*, *54*, 358 (1973).

220. L. J. Lis, J. W. Kaufman, and D. F. Shriver, *Biochim. Biophys. Acta*, *406*, 453 (1975).

221. R. Faiman and D. A. Long, *J. Raman Spectrosc.*, *3*, 379 (1975).

222. R. Mendelsohn, *Biochim. Biophys. Acta*, *290*, 15 (1972).

223. R. F. Schaufele and T. Shimanouchi, *J. Chem. Phys.*, *47*, 3605 (1967).

224. H. Akutsu and Y. Kyogoku, *Chem. Phys. Lipids*, *14*, 113 (1975).

225. B. Abramson, W. T. Norton, and R. Katzman, *J. Biol. Chem.*, *240*, 2389 (1965).

226. D. Chapman and A. Morrison, *J. Biol. Chem.*, *241*, 5044 (1966).

227. B. J. Bulkin and N. I. Krishnamachari, *Biochim. Biophys. Acta*, *211*, 592 (1970).

228. H. Akutsu, Y. Kyogoku, H. Nakahara, and K. Fukuda, *Chem. Phys. Lipids*, *15*, 22 (1975).

229. H. Müller, S. Friberg, and H. H. Bruun, *Acta Chem. Scand.*, *23*, 515 (1969).

230. T. Takenaka, K. Nogami, H. Gotoh, and R. Gotoh, *J. Colloid Interfac. Sci.*, *35*, 395 (1971).

231. R. N. Jones and C. Sandorfy, in *Technique of Organic Chemistry* (A. Weissberger, ed.), Vol. 9, Wiley-Interscience, New York, 1956, p. 247.

232. L. F. Fieser and M. Fieser, *Steroids*, Reinhold, New York, 1959.

233. E. L. Eliel, N. L. Allinger, S. J. Angyal, and G. A. Morrison, *Conformational Analysis*, Wiley-Interscience, New York, 1965, p. 256.

234. K. Dobringer, E. R. Katzenellenbogen, and R. N. Jones, *Infrared Absorption Spectra of Steroids - An Atlas*, Vol. 1, Wiley-Interscience, New York, 1958.

235. G. Roberts, B. S. Gallagher, and R. N. Jones, *Infrared Absorption Spectra of Steroids - An Atlas*, Vol. 2, Wiley-Interscience, New York, 1958.

236. J. E. Zull, S. Greanoff, and H. K. Adam, *Biochemistry*, *7*, 4172 (1968).

237. F. S. Parker and K. R. Bhaskar, *Biochemistry*, *7*, 1286 (1968).

238. A. H. Maddy and B. R. Malcolm, *Science*, *150*, 1616 (1965).

239. D. Chapman, V. B. Kamat, and R. J. Levene, *Science*, *160*, 314 (1968).

240. D. F. H. Wallach and P. H. Zahler, *Proc. Natl. Acad. Sci. U.S.A.*, *56*, 1552 (1966).

241. T. J. Jenkinson, V. B. Kamat, and D. Chapman, *Biochim. Biophys. Acta*, *183*, 427 (1969).

242. D. H. Green and M. R. J. Salton, *Biochim. Biophys. Acta*, *211*, 139 (1970).

243. A. S. Gordon, J. M. Graham, B. R. Fernbach, and D. F. H. Wallach, *Federation Proc.*, *28*, 404 (1969).

244. G. L. Choulos and R. F. Bjorklund, *Biochemistry, 9,* 4759 (1970).

245. J. Avruch and D. F. H. Wallach, *Biochim. Biophys. Acta, 241,* 249 (1971).

246. R. J. Cherry, K. U. Berger, and D. Chapman, *Biochem. Biophys. Res. Commun., 44,* 644 (1971).

247. J. M. Graham and D. F. H. Wallach, *Biochim. Biophys. Acta, 241,* 180 (1971).

248. K. Nakamura, D. N. Ostrovsky, T. Miyazawa, and S. Mizushima, *Biochim. Biophys. Acta, 332,* 329 (1974).

249. B. J. Bulkin, *Biochim. Biophys. Acta, 274,* 649 (1972).

250. D. F. H. Wallach and S. P. Verma, *Biochim. Biophys. Acta, 382,* 542 (1975).

251. S. P. Verma and D. F. H. Wallach, *Biochim. Biophys. Acta, 401,* 168 (1975).

252. J. L. Lippert, L. E. Gorczyca, and G. Meiklejohn, *Biochim. Biophys. Acta, 382,* 51 (1975).

253. A. T. Tu, B. Prescott, C. H. Chou, and G. J. Thomas, Jr., *Biochim. Biophys. Res. Commun., 68,* 1139 (1976).

254. H. Akutsu, K. Uehara, T. Shinpo, Y. Kyogoku, and Y. Akamatsu, unpublished.

255. T. Kitagawa, Y. Kyogoku, T. Iizuka, and M. Saito, *J. Amer. Chem. Soc., 98,* 5169 (1975).

256. R. H. Felton and N. T. Yu, in *The Porphyrins* (J. Dorphin, ed.), Academic Press, New York, 1976.

257. G. Placzek, in *Rayleigh and Raman Scattering,* UCRL Trans No. 526L from *Handbuch der Radiologie* (E. Marx, ed.), Vol. 2, p. 209, Akademische Verlagsgesellshaft VI., Leipzig, 1934.

258. B. Prescott, C. H. Chou, and G. J. Thomas, Jr., *J. Phys. Chem., 80,* 1164 (1976).

259. M. N. Siamwiza, R. C. Lord, M. C. Chen, T. Takamatsu, I. Harada, H. Matsuura, and T. Shimanouchi, *Biochemistry, 14,* 4870 (1975).

260. M. Tasumi, T. Shimanouchi, and T. Miyazawa, *J. Mol. Spectrosc., 9,* 261 (1962).

Chapter 12

POLYMERS

Sandra C. Brown*

Arapahoe Chemicals, Inc.
Boulder, Colorado

and

Albert B. Harvey

Chemistry Division--Code 6110
Naval Research Laboratory
Washington, D.C.

*Present address: Food and Drug Administration, Bureau of Drugs,
Division of Cardio-Renal Drug Products, Rockville, Maryland.

I. INTRODUCTION

As is the case with any class of materials that enjoys widespread
use in varied applications, the properties of polymeric materials
have been the subject of numerous studies. This is especially true
of their IR and Raman spectroscopic properties [1-7]. The spectroscopic data that have been collected have given important insights
into polymeric properties.

IR spectroscopy masked Raman methods in the vibrational study
of polymers until very recently, mainly because of the differences
in sample-handling techniques and instrumentation. It is a tribute
to the laser that this chapter covers IR *and* Raman spectroscopy of
polymers. Only recently has the spectroscopic information regarding
polymers been expanded to include significant results using Raman
techniques [8]; less than a dozen polymeric materials had been
studied prior to development of laser Raman spectroscopy in 1966 [4].
Studies of IR and Raman spectra of polymers now include the following:
(1) spectra of reference compounds used for identification of unknown samples by comparison, (2) studies of the normal vibrations of
polymeric systems, (3) quantitative applications, (4) studies of the
changes which take place during curing and vulcanization or other
chemical changes, (5) conformational studies which help elucidate the
transitions which take place between different structures and shed
light on the structures themselves, and (6) miscellaneous studies,
such as those of bonding in polymer and copolymer systems using
vibrational information to determine the types of linkages that are
formed.

Because of the wide variety of polymeric materials, the spectroscopic data in the literature are quite extensive. Studies are
often available for various compounded systems, pyrolysis products,

and degradation products. As an approach to discussion of the above,
we shall first look at the techniques used in studying polymeric
systems. The results obtained for the systems studied can then be
tied to the methods used.

II. EXPERIMENTAL TECHNIQUES

A. Infrared

Polymeric systems have been studied by IR spectroscopy for a
number of years. The use of conventional IR techniques [9] has usually required that at least some sample preparation be carried out prior
to obtaining the spectrum, since the physical state of most polymers
makes them unsatisfactory samples for direct placement in the IR
beam. Sample preparation methods include using solutions, casting
thin films on IR transmitting plates, preparing melt samples between
plates, mulls, pelletizing, and more recently, attenuated total reflectance (ATR). Pure polymers are more easily dealt with and the
spectra are generally more simple to analyze than are compounded
systems.

A summary of the usual alternatives available for obtaining the
IR spectrum of a polymeric sample is given in Fig. 1. Pure samples
are easier to work with because compounding and curing not only add
small amounts of other substances, such as plasticizers, fillers, and
antioxidants, but also increase the problem of getting the sample into
a suitable form by solution, grinding, etc. Some pure polymers which
are powders can be examined as KBr pellets or Nujol mulls. This is
probably the simplest method of preparation but not always the most
satisfactory, because differences in refractive index between samples
and suspending medium may give rise to poor-quality spectra. A
method for getting some polymers into a powdered form suitable for
preparing KBr pellets or mulls has been reported [10]. Poly(oxymethylene) was dissolved in dimethyl sulfoxide (DSMO) and precipitated
by adding methanol or ethanol. After washing a few times to remove
the DMSO and vacuum drying to remove the alcohol, the dry powder

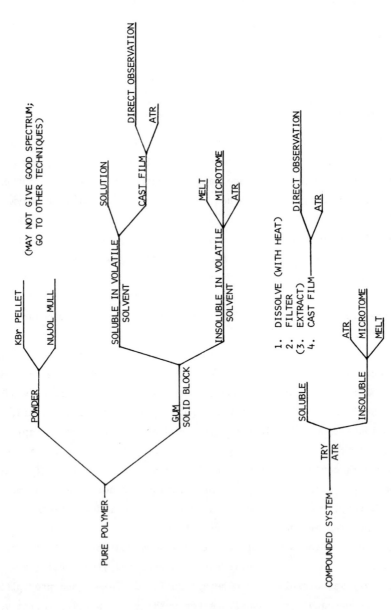

FIG. 1. Techniques useful for preparing polymer samples for examination by IR spectroscopy.

could be prepared for recording the spectrum. Good-quality spectra can be obtained from thin films of polymers which are cast either on spectral plates or on glass or some other substance from which they can be peeled and then mounted in the IR beam. This is a very satis-factory method for samples that are soluble in a solvent that is relatively volatile. Polystyrene films frequently used for calibra-tion of IR spectrophotometers are familiar examples of a polymer sample prepared as a thin film. The main problems encountered are the formation of films of the right thickness (not too thick so that bands are off scale and not too thin so that bands are not sufficient-ly intense) and the affirmation that no solvent remains trapped in the film to give spurious bands in the spectrum. The former problem is usually minimized with a little experience, and a vacuum desiccator solves the latter. Samples which are not soluble in suitable solvents can be handled by melting a small amount of the material between two plates. Again, the film can be used in this form or removed and mounted in some other way. For the very stubborn samples which do not submit to any of the above techniques (this includes many polymers), the ATR technique is frequently useful [11]. Commercial units (Fig. 2) of several different manufactures are available, and instructions will indicate methods of handling samples such as polymers. There is no sample preparation required other than obtaining a flat portion of sample to place against the ATR prism. Polymer samples are often deformable enough that this presents no real problem. Sample thick-ness, color, and nonuniformity cause no special obstacles for ATR. However, it should be noted that in ATR one is viewing only the first few micrometers of surface and the results do not necessarily suggest the composition of the bulk (see Fig. 3). Of course, this particular characteristic can be turned to advantage for determining surface effects, oxidation, etc.

Compounded systems provide fewer alternatives for methods of preparation. The methods of casting films and ATR are also applicable to these systems, but with slightly more difficulty. It is often necessary to filter solutions of these materials to remove fillers or other materials added in compounding. The problem is greater if

FIG. 2.   ATR attachment for IR spectrophometer.   (Photo courtesy of
Barnes Engineering Company).

heating is necessary to get the polymer into solution, but it usually
can be handled.   For those samples which will not go into solution or
for which a nondestructive method is desired, ATR offers a suitable
alternative.

In addition to these general techniques which are applicable to
all types of samples (not just polymers), a few specialized methods
have also been reported recently.   Descriptions of two types of cells
have been given for obtaining spectra of polymers at high temperatures
[12,13].   Feairheller and Crawford [12] reported a cell which could
be used for studying polymers at temperatures up to 600°C in nitrogen
or argon atmospheres, or in a vacuum.   The windows were removable for
ease of cleaning.   An attachment devised by Panov and Zheleznov [13]

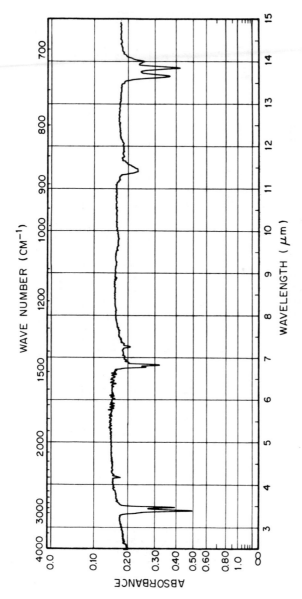

FIG. 3(a). ATR spectrum of wax coating on a natural rubber sample. (Courtesy of Barnes Engineering Company.)

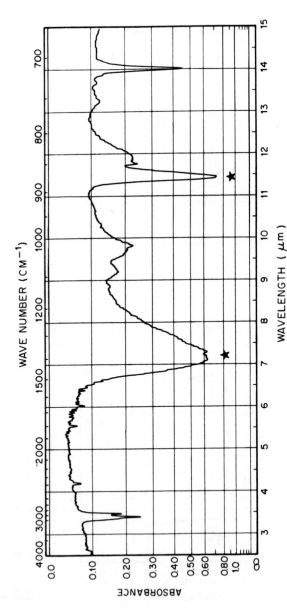

FIG. 3(b). ATR spectrum of the same natural rubber sample as in Fig. 3(a) after the wax was scraped off (*indicates absorption due to calcium carbonate filler). (Courtesy of Barnes Engineering Company.)

consisted of an evacuated chamber containing heating elements in
which polymer samples could be heated up to 200°C and stretched up
to 500%. Another device, reported by Burley and Woollerton [14],
was used for spinning samples to reduce inhomogeneity effects in
quantitative IR absorbance measurements. Application of the device
was illustrated for samples of polyethylene and polypropylene. Another type of high-temperature cell was used for obtaining IR emission
spectra of polymers by melting samples onto the surface of a bronze
sinter disk [15]. Pore sizes from 40 to 110 μm were used, and improvement of spectrum quality was found with increasing pore size.
Optimum temperatures found were 180°C for polyethylene and polypropylene and 250°C for polystyrene.

B. Raman

The discussion of sample preparation techniques for Raman spectroscopy of polymeric materials is much simpler, in large part as a
result of the use of the laser source in current instrumentation.
Until this improvement in instrumentation was made available, sample
preparation was quite tedious and involved much purification of
samples to remove even minute amounts of impurities, which caused
"fluorescence" problems in obtaining spectra. Samples which were
red or yellow could not be studied at all, because the very common
blue line of the mercury arc source was totally absorbed. Raman
spectral studies for such simple polymers as polyethylene and polystyrene were the only ones performed before the coming of the laser.
A review of the problem and meager amount of information concerning
the Raman spectra of polymers up until 1964 may be found in an article
by Nielsen [16]. The fact that a wider range of samples can now be
handled is reflected in the number of publications in recent years
concerned with polymers, most of which had not previously been studied
by Raman spectroscopy.

Because of the laser, very little sample preparation is now
necessary to obtain at least a survey spectrum of a polymer, and sample
size and thickness present no problem as they do in IR spectroscopy.

The only limitation on size is that the sample fit into the space
provided in the instrument. Color is no longer the problem it was
formerly. Red and yellow samples are easily studied with the use of
632.8 nm radiation of the He-Ne laser or the Kr^+ 647.1-nm laser line,
and almost any colored sample can be examined with the wide range of
exciting lines that are now available in tunable lasers. The main
problem which remains in handling polymeric samples and biological
samples alike is sample fluorescence. This phenomenon is not yet
fully understood but is fairly widespread in Raman spectroscopy. It
seems to be caused by impurities in samples, because purification
often decreases the amount of background fluorescence. The fluorescence is manifested in the spectrum by very strong general background
emission, which is often much greater than the desired, but weaker,
Raman scattering. Figure 4 shows a spectrum of a polymeric sample
which exhibits this problem.

There appear to be several methods for reducing or eliminating
the source of aggravation. One method already mentioned involves
the careful purification of the material. This treatment is usually
very tedious, often unsatisfactory, and impossible for many materials.
Some success can be achieved by shifting to different exciting frequencies, e.g., He-Ne or Ar^+ lasers. Still another method is to
irradiate the sample in the laser beam for several hours, even days.
After this period of time some materials show markedly improved spectra (see Fig. 5). Finally, a much more elegant method for eliminating
this bothersome background emission is presently emerging. The idea
is not a new one and can be described in very simple terms. Fluorescence, unlike Raman scattering, is not an "instantaneous" process and
hence can be reduced by time discrimination. Raman scattering occurs
on about the 10^{-13}- to 10^{-14}-sec scale whereas most fluorescence events
are delayed by picoseconds to microseconds. Preliminary experiments
at Block Engineering, Inc. [17] now suggest that in some instances,
at least, the use of pulsed lasers (subnanosecond) and fast-response
photomultiplier tubes can reduce background emission by one to two
orders of magnitude at a small cost in Raman signal.

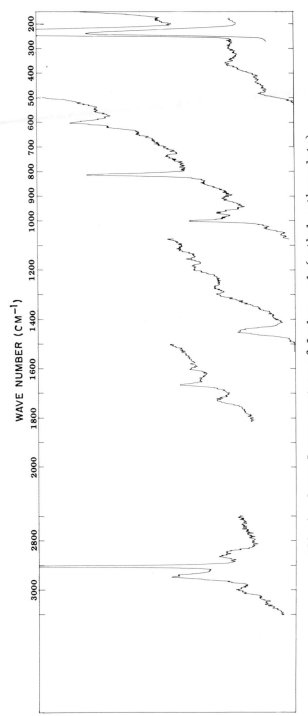

FIG. 4. Laser Raman spectrum of Implex, poly(methyl methacrylate).

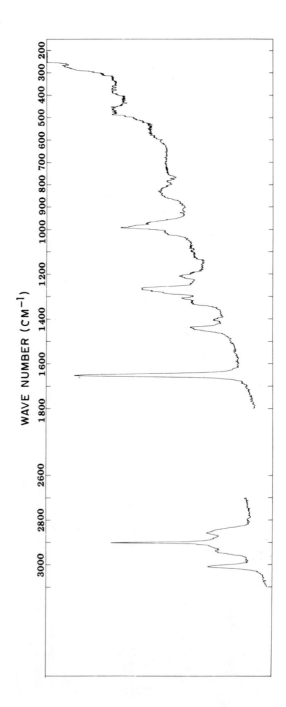

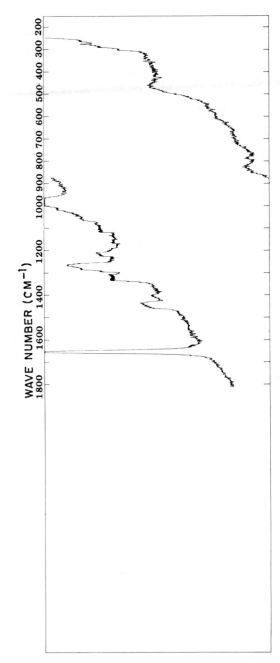

FIG. 5. Laser Raman spectra of Ameripol CB 220, polybutadiene. Lower: immediately after sample was placed in position; upper: after sample was in beam for 2 hr.

Sampling for Raman spectroscopy is generally a nondestructive method, except in cases where sample treatment (e.g., purification) is necessary or where the sample absorbs the laser light sufficiently to decompose the material. This offers advantages to those who wish to observe systems in the state that they will be eventually used and affords the obvious simplification in procedure. Another useful benefit from Raman spectroscopy is that water is only a weak Raman scatterer while it is a nearly ubiquitous absorber in the IR. This allows systems to be studied in water solution or some sort of controlled humidity atmosphere with little interference from water. This is of particular importance to polymer systems of biological interest.

Various types of cells are available for Raman instruments, but are usually not necessary for polymers unless solutions or liquid polymers are being examined. Because Raman is a scattering and not an absorption phenomenon, too much sample can almost never be a problem--strong peaks are just attenuated to bring them on scale. Samples are singly mounted in the beam and the forward scattering is recorded, or the sample surface is mounted in the beam at some angle near 90° to the normal and right-angle scattering is observed.

III. COMPARISON OF UTILITY OF INFRARED AND RAMAN SPECTROSCOPY

Other chapters deal with some comparisons of IR and Raman spectra, but it may be useful to discuss here some of the pertinent comparisons as they apply to polymers.

One area of interest in the study of polymers is identification of components which sometimes are present in only 5 to 10% concentrations in copolymer systems. Studies of various types of degradation are also concerned with observing changes which affect only a small part of the entire polymeric material. In such cases, it is important to have a tool which is sensitive to the particular functional group which reflects the change of interest. The dependence of Raman scatter on changes in polarizability makes it quite sensitive to

vibrations that are of a symmetrical nature and to vibrations involving larger atoms. IR absorption, on the other hand, resulting
from interaction of radiation with vibrations which involve a change
in dipole moment is more useful in the study of vibrations of unsym-
metrical groups and those which involve atoms of very different
electronegativities. Some types of vibrations which are found to
give rise to strong bands in IR absorption, and conversely weak lines
in Raman spectra, are the stretches of the carbonyl group, ethers,
nitriles, O-Si-O groups (antisymmetric), and OH. Raman spectra usually are better indicators than IR for C=C, C≡C, phenyl, S-S, and C-S
groups. A comparison of some of these differences is given in Table 1.
Advantages of these differences are seen in the applications mentioned
above. For example, vulcanized samples show very strong S-S and C-S
bands in the Raman spectrum which are not observed in the IR [18].
Moreover, in the study of photodegradative and thermal oxidation of
polymers, changes most likely to occur are the formation of C=O, O-H,
and C-O bonds. These are produced in small amounts in the early
stages of degradation and would be detectable in an IR spectrum much
earlier than in a Raman spectrum. Examples of this utility are displayed in the spectra in Fig. 6. Raman spectroscopy holds the advantage over IR in the detection of small amounts of vinyl and phenyl
groups copolymerized in silicone systems. The IR and Raman spectra
compared in Fig. 7 show the difference in band intensities observed
in the two spectra for the same compound. Other examples of differences between IR and Raman spectra of some polymer systems and an
introduction to Raman group frequencies are given by Sloane [20].

Another major area of difference between IR and Raman spectroscopy of polymers lies in the sampling techniques required. This is
probably more of a difference for polymers than for most other types
of samples because of their unique properties. Preparation of the
sample itself has already been discussed. In addition, there are
major differences in the instrumentation which are important considerations. The weak spectrum of water in the Raman effect makes
purging of the instrument with dry air or nitrogen unnecessary, unlike
IR spectroscopy. Therefore, the time required for equilibration of the

TABLE 1

Common Intensity Differences Between IR and Raman Spectra

Strongly Raman active		Strongly IR active		Strong in both	
Vibration	Frequency range (cm⁻¹)	Vibration	Frequency range (cm⁻¹)	Vibration	Frequency range (cm⁻¹)
Aromatic C-H stretch	3000-3100	C=O Stretch	1600-1800	Aliphatic C-H stretch	2800-3000
C=C Stretch	1600-1700	C-O Stretch	900-1300	C≡N Stretch*	2200-2300
C≡C Stretch	2100-2250	O-H Stretch (H bonded)	3000-3400	Si-H Stretch	2100-2300
S=S Stretch (Se-Se, etc.)	400-500	Aromatic CH out-of-plane deformation	650-850	C-Halogen stretch	500-1400
C-S Stretch (C-Se, etc.)	600-700	N-H Stretch (H bonded)	3100-3300		
Aromatic ring breathing	950-1050	Si-O-Si Antisymmetric stretch	1000-1100		
Aromatic C=C in-plane vibration	1500-1700				
N=N Stretch	1575-1630				
Si-O-Si Symmetrical stretch	450-550				

In general:

1. Vibrations of symmetrical groups

2. Stretching of bonds, particularly nonpolar or slightly polar bonds

In general:

1. Vibrations of asymmetric groups

2. Bending modes

3. Stretching of polar bonds

*See Chap. 6, Sec. III.E.

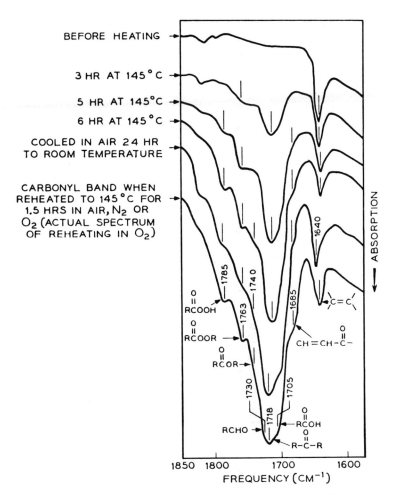

FIG. 6. Carbonyl stretching region of Marlex 50 polyethylene during and after heat oxidation in air at 145°C (reprinted from Ref. 19, by permission of John Wiley and Sons, Inc.).

instrument atmosphere after opening the sample compartment or the main body of the instrument is a factor only in IR spectroscopy. Raman spectroscopy, on the other hand, often necessitates some skill on the part of the operator, particularly in the proper placement of liquid samples in the laser beam. Some manipulation is usually required before the optimum scattering is obtained. One of the major advantages of Raman spectroscopy is the availability of the entire vibrational spectrum on one instrument. A separate, far-IR spectrophotometer is

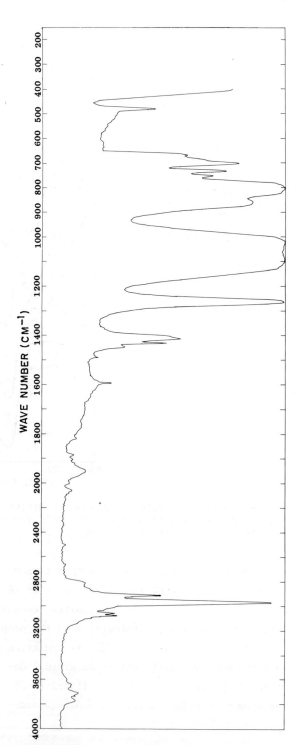

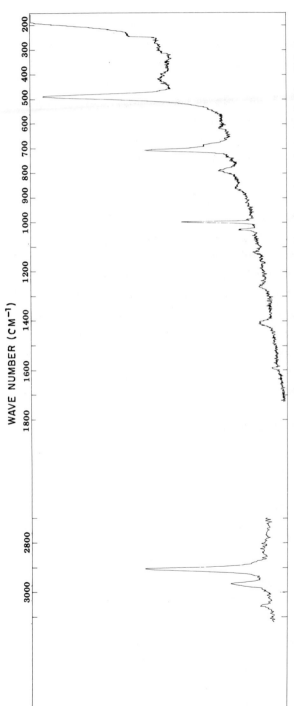

FIG. 7. IR (upper) and laser Raman (lower) spectra of Silastic 440, a mixed methyl phenyl vinyl silicone polymer.

required for information in the region below 300 cm^{-1} because the
conventional IR systems do not cover this range. Raman scattering,
however, is observed as a difference between the frequency of the
exciting line and the vibrational frequency, so that frequencies for
the entire range are as accessible as the capability of approaching
the exciting line. By narrowing slits, using a triple monochromator,
etc., vibrations lower than 50 cm^{-1} can be observed with little difficulty. The operation of Raman instruments in the visible spectral
range also makes a difference in the substances that may be used as
window and cell materials. Glass becomes the material of choice in
Raman spectroscopy, and there is no polishing of plates, which is
commonplace in IR spectroscopy. Ordinary capillary tubes serve quite
well as Raman cells for liquid samples and can be readily sealed and
modified for temperature studies either by heating or cooling.

IR and Raman spectroscopy are complementary techniques. Neither
can give all the information that the two techniques can provide together because of differences in selection rules. Depending on the
type of information that is required from a particular study, however,
one or the other of the techniques may prove better suited because of
sensitivity required or sampling methods that can be used.

IV. APPLICATIONS OF VIBRATIONAL SPECTROSCOPY

A. Use of Reference Spectra
for Identification Purposes

1. *General Discussion*

Extensive use has been made of IR spectroscopy as an analytical
tool in identification of polymers. Collections of spectra of plastics and rubbers are available [21-26] as well as compilations of
reference spectra for plasticizers, stabilizers, and other additives
[6,27]. Use of these catalogs greatly simplifies the identification
of an unknown polymer system. Figure 8 reproduces the guide prepared
by Cain and Stimler [23] for the classification of polymeric materials
from their IR spectra. Most general classes of plastics and rubbers

are included in this scheme, and brand names are given for the particular polymers studied in their reports. An unknown sample can be
classed as to type and identification made by comparison of its spectrum with spectra of known samples.

This use of IR spectra is based upon the application of group
frequency correlations to polymer molecules in essentially the same
manner as for small molecules. It is obvious from examination of
polymer spectra that they are not nearly as complicated as might be
expected from a naïve consideration of 3N-6 possible normal vibrations. The relative simplicity of the spectra is due to the coincidence of so many of the vibrations created by the repeat units. The
same group frequency correlations that have been useful for simple
molecules can, therefore, be readily applied to polymers [28]. Similar application of Raman spectroscopy is possible, utilizing its
particular sensitivities to certain functional groups, but the actual
application is much more limited than that of IR [8,29,30]. This is
mainly a result of the "late blooming" of Raman as applied to polymers
in general and because the availability of Raman instruments is much
less (the cost of instrumentation is greater) and the demand for such
information much smaller. Some work has been attempted in the direction of compiling reference spectra of polymers, and this can be used
in discussing the utilization of group frequencies in the identification of polymers [20].

Because of the difficulty in preparing some vulcanized rubber
samples for IR spectroscopy, studies have also been made of pyrolysis
products of such systems for identification purposes [26,31-35]. The
pyrolysis technique is intended for those samples which cannot be
studied by other sampling techniques due to such characteristics as
insolubility, infusibility, or toughness. The spectra obtained from
pyrolyzates of such polymers, while usually different from those of
the materials themselves, are useful in the identification of the
material by comparison with reference spectra prepared in the same
manner. The pyrolyzate spectra are generally similar to the spectra
of the polymer because many polymers tend to pyrolyze into monomeric

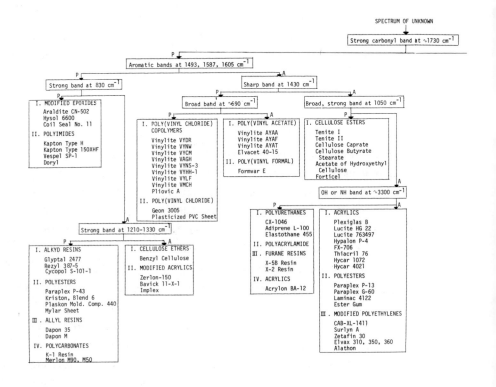

FIG. 8. Scheme for classification of polymeric materials from their IR spectra. P = present; A = absent.

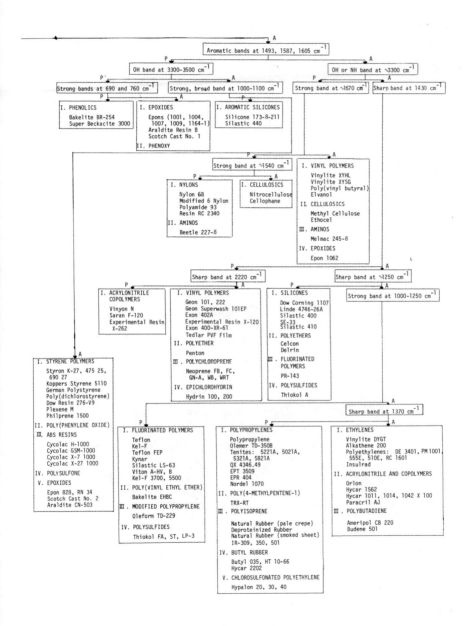

or low-molecular weight components. Some special types of cells
have been described for obtaining IR spectra of rubber pyrolyzates
[31], and special ATR attachments are now commercially available,
made particularly for this purpose [36]. The technique is generally
one of separating the rubber material from the fillers and other
agents that are added in curing and which confuse the spectra, and
then collecting the desired material in a suitable form for recording
the spectrum, usually as a gas or liquid. Some attempts have been
made at cataloging pyrolyzate reference spectra [26,33].

The Barnes commercial apparatus (see Fig. 9) differs from most
techniques described earlier in that it allows for examination of
the gas phase pyrolysis products as well as liquid condensates. This

FIG. 9. Pyrolysis unit for collection of samples for examination by
IR spectroscopy. Cells are for (left) liquid condensate and (right)
gaseous products. (Figure reproduced by courtesy of Barnes Engineering Company.)

provides the opportunity for examination of additional information
that previously was not available. Cassels [35] has reported ATR
spectra of a number of polymers which include the starting materials,
gas phase pyrolysis products, and condensed phase pyrolyzates. The
vapors produced were usually indicative of the polymers from which
they came, although they were not necessarily monomer species. For
example, HCN was always produced from polyacrylonitrile or acrylo-
nitrile-containing copolymers, indicative of the acrylonitrile seg-
ment of the molecule. Acetates produced acetic acid; polyamides,
ammonia; and polyesters, acetaldehyde (see Fig. 10). It was suggested
that the vapor phase pyrolysis products could be quite useful in distinguishing between polymers which showed few differences in their
condensed phase spectra.

The application of the pyrolysis technique has been confined
mainly to combinations with IR spectroscopy or gas chromatography.
There is no reason, however, why a similar combination with Raman
spectroscopy should not be possible, although such destruction of the
sample is less often necessary in Raman scatter than IR absorption.
Should such information be desirable, the main adaptation of the
technique that would be required would involve proper collection of
the pyrolyzate sample in a cell for obtaining the Raman spectrum.

Some typical results of spectroscopic analysis of polymers can
be summarized here, using the identification of Fig. 8 as a basis
for discussion. This scheme was originally prepared by Kagarise and
Weinberger [21] and was modified and expanded in later works [22,23].
The classification which has been made is based on characteristics
observed in the IR spectra. Some attempt will be made here to point
out utilities of Raman spectra in this application. Table 2 contains
structural information for most of the polymer types discussed in
the following sections.

2. *Polymers Containing Carbonyl and Aromatic Absorption*
The presence or absence of a strong C=O absorption in the IR
spectrum in the 1700- to 1750-cm^{-1} range roughly divides polymer types
into two groups. The C=O group may also show a Raman line but generally it would be observed to be less intense than the IR counterpart.

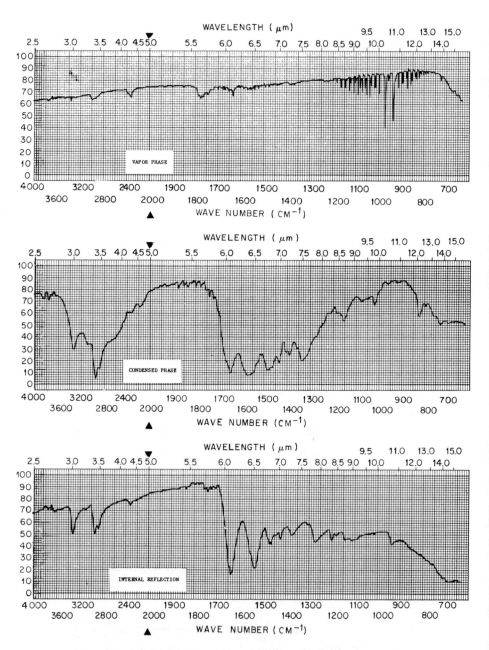

FIG. 10. IR spectra obtained from sample of Nylon 66. Upper, gaseous pyrolysis products; middle, condensed phase pyrolysis products; and lower, ATR spectrum of bulk material. (Photo reproduced by courtesy of Wilks Scientific Corporation.)

TABLE 2

Structures of Various Polymer Types

Type	Typical structure
Acrylic	$\{CH_2-\underset{\underset{\underset{O-CH_3}{O}}{\overset{\overset{CH_3}{\vert}}{\underset{\vert}{C}}}}\}_n$
Alkyd (polybasic acid + polyhydric alcohol)	(structure with benzene ring, $-\underset{\overset{\Vert}{O}}{C}$, $\underset{\overset{\Vert}{O}}{C}-OCH_2-\underset{\underset{O=C-C_{17}H_{31}}{\overset{\vert}{O}}}{CH}-CH_2-O-)_n$
Butyl rubber (isobutylene-isoprene)	$\{\underset{\underset{CH_3}{\vert}}{\overset{\overset{CH_3}{\vert}}{C}}-CH_2\}_n$ $\{CH_2-\underset{\underset{CH_3}{\vert}}{C}=CH-CH_2\}_m$
Cellulose ester	(cyclic structure with OH, X, CH-CH, CH-O, H$_2$CY)
Cellulose ether	(cyclic structure with OH, OR, CH-CH, CH-O, H$_2$C-OR)
Chloroprene	$\{CH_2-\underset{\underset{Cl}{\vert}}{C}=CH-CH_2\}_n$
Chlorosulfonated polyethylene	$\{(CH_2-CH_2-CH_2-\underset{\underset{Cl}{\vert}}{CH}-CH_2-CH_2-CH_2)_x\underset{\underset{\underset{Cl}{\vert}}{\overset{\overset{SO_2}{\vert}}{CH}}}\}_n$

TABLE 2 (continued)

Type	Typical structure
Epichlorohydrin	$\{CH\text{-}CH_2\text{-}O\}_n$ with CH_2Cl
Epoxide	$H_2C\!-\!CH\text{-}CH_2\{O\text{-}\emptyset\text{-}\underset{CH_3}{\overset{CH_3}{C}}\text{-}\emptyset\text{-}O\text{-}CH_2\text{-}CH\text{-}CH_2\}_n\text{-}O\text{-}\emptyset\text{-}\underset{CH_3}{\overset{CH_3}{C}}\text{-}\emptyset\text{-}O\text{-}CH_2\text{-}CH\!-\!CH_2$ (with OH)
Fluorosilicone	$\{Si\text{-}O\}_n$ with CH_xF_{3-x} above and below
Isoprene rubber	$\{CH_2\text{-}C\!=\!CH\text{-}CH_2\}_n$ with CH_3
Nylon (polyamide)	$\{\overset{O}{\overset{\|}{C}}\text{-}(CH_2)_4\text{-}\overset{O}{\overset{\|}{C}}\text{-}\underset{H}{N}\text{-}(CH_2)_6\text{-}NH\}_n$
Phenolic (phenolic-aldehyde resin)	phenol ring with OH, $\{\text{-}CH_2\text{-}\}_n$
Phenoxy	$\{\emptyset\text{-}\underset{CH_3}{\overset{CH_3}{C}}\text{-}\emptyset\text{-}O\text{-}CH_2\text{-}CH\text{-}CH\text{-}\}_n$ with OH OH
Polyacrylamide	$\{CH_2\text{-}CH\}_n$ with $C\!=\!O$, NH_2

TABLE 2 (continued)

Type	Typical structure
Polyacrylonitrile	$\{CH_2CH\}_n$ with $C{\equiv}N$
Polybutadiene	$\{CH_2-CH{=}CH-CH_2\}_n$
Polycarbonate	
Polyester (dibasic acid + dihydric alcohol)	
Polyether	$\{CH_2-O-CH_2\}_n$
Polyethylene	$\{CH_2-CH_2\}_n$
Polyimide	
Poly(4-methylpentene-1)	$\{CH_2-CH\}_n$ with CH_2, $CH(CH_3)_2$
Poly(phenylene oxide)	

TABLE 2 (continued)

Type	Typical structure
Polypropylene	$\{CH_2-\underset{\underset{CH_3}{\mid}}{CH}\}_n$
Polysulfide	$\{CH_2-CH_2-S_x\}_n$ $\{CH_2-CH_2-S_x\}_m$ $\{CH_2-CH_2-O-CH_2-O-CH_2-CH_2-S_x\}_n$
Polysulfone	
Polystyrene	
Polyurethane	$\{R-\underset{\underset{R'}{\mid}}{N}-\underset{\overset{\parallel}{O}}{C}-O-R''\}_n$
Poly(vinyl acetal) Poly(vinyl alcohol) condensed with aldehyde	
Poly(vinyl acetate)	
Poly(vinyl alcohol)	$\{CH_2-\underset{\underset{OH}{\mid}}{CH}\}_n$

TABLE 2 (continued)

Type	Typical structure
Poly(vinyl chloride)	$\begin{matrix} \{CH_2\text{-}CH\}_n \\ \quad\;\; \vert \\ \quad\;\; Cl \end{matrix}$
Poly(vinyl ethyl ether)	$\begin{matrix} \{CH_2\text{-}CH\}_n \\ \qquad \vert \\ \quad\; O\text{-}C_2H_5 \end{matrix}$
Poly(vinylidene fluoride)	$\{CH_2\text{-}CF_2\}_n$
Silicone (polysiloxane)	$\begin{matrix} R \\ \vert \\ \{Si\text{-}O\}_n \\ \vert \\ R \end{matrix}$
Teflon (poly(tetrafluoroethylene))	$\{CF_2\text{-}CF_2\}_n$

A check for phenyl groups provides a further subdivision. IR bands at 1493, 1587, and 1605 cm^{-1}, characteristic of benzene rings, give a good indication of this functionality but are not particularly strong vibrations. A useful feature of the Raman spectrum for compounds containing phenyl groups is the ring-breathing mode near 1000 cm^{-1}. It is a very strong, sharp line observed for monosubstituted, meta-disubstituted, and 1,3,5-trisubstituted rings. The line is observed at 1050 cm^{-1} in ortho-disubstitution and is absent from spectra of para-disubstituted compounds. Another vibration helpful in the diagnosis of presence of phenyl groups is the aromatic C-H stretch (3000 to 3100 cm^{-1}), generally stronger in Raman than IR. The number of polymers containing both a carbonyl group and a phenyl ring is relatively few. The additional presence of a strong IR band at 830 cm^{-1} splits out two groups of polymers, the modified epoxides and polyimides. Those spectra which did not contain the 830-cm^{-1} band are examined for a strong band in the 1210- to 1330-cm^{-1} region.

This characterizes the ester group of compounds, such as alkyd resins
of the phthalic ester type, polyesters, and polycarbonates. This
vibration will be a more prominent feature of the IR than the Raman
spectrum. The group remaining, which did not exhibit the 1210- to
1330-cm^{-1} band, includes cellulose ethers and modified acrylics.
The polymers in the group are somewhat specialized cases of groups,
most of the other members of which are found elsewhere in the classification scheme. Benzyl cellulose appears here because of its phenyl
group content whereas most of the other cellulose ethers are found
later in the scheme. The modified acrylics contain copolymerized
styrene and thus fall in this group.

3. Polymers Containing Carbonyl Absorption

The next large group in the classification scheme contains those
materials which exhibit C=O absorption but do not contain aromatic
groups. A sharp absorption band at 1430 cm^{-1} arising from a C-H
bending motion characterizes vinyl polymers, which can be further
subdivided by the indication of a broad 690-cm^{-1} band. The polymers
containing this band (the C-Cl stretch) are poly(vinyl chloride)
(PVC) systems, either copolymers with carbonyl-containing material
or PVC itself which has been modified in some way. The C-Cl stretch
also produces strong Raman scatter, and this Raman frequency can be
used to distinguish between PVC and poly(vinyl acetate) systems in
the same manner. Those compounds which do not contain the 690-cm^{-1}
absorption at this point in the scheme are primarily poly(vinyl
acetates) and poly(vinyl acetals). Going to the other side of this
system to the compounds not exhibiting the sharp 1430-cm^{-1} vibration,
examination of the spectrum for strong, broad absorption at 1050 cm^{-1}
distinguishes between cellulose esters and remaining members of the
group. This vibration is characteristic of the cellulose ring [21].
In the remaining group, OH or NH stretching absorption near 3300 cm^{-1}
characterizes a group of four polymer types, and three polymer types
remain which do not show this absorption. The polyurethanes, poly-
acrylamide, furane resins, and an acrylic (acrylic ester-acryloni-
trile copolymer) exhibit the 3300-cm^{-1} absorption. This vibration

will not be useful in Raman spectra, as NH and OH stretches produce
characteristically weak scatter. The group which does not exhibit
NH or OH absorption contains most of the acrylics, some polyesters,
and a few modified polyethylenes.

4. Polymers Containing Aromatic Absorption

Going back to the first main subdivision now, the second major
group of non-carbonyl-containing polymers is divided by the presence
of aromatic groups in the same manner as before. The presence of an
OH group (3300- to 3500-cm^{-1} absorption) in addition to the phenyl
group characterizes phenolic, epoxide, and phenoxy polymers. The
phenolic group is differentiated from the rest by the presence of
strong 690- and 760-cm^{-1} bands, characteristic of monosubstituted
benzene rings. Epoxide and phenoxy systems contain disubstituted
rings. As mentioned in the earlier discussion, Raman spectra can be
useful in differentiating between various types of aromatic substitution. Of those remaining polymers which do not contain OH groups,
the aromatic silicones are characterized by strong, broad absorption
in the 1000- to 1100-cm^{-1} range. This arises from the Si-O-Si antisymmetric stretching vibration. An interesting comparison is noted
in observing the Raman spectra of these compounds. There is no strong
band in the 1000- to 1100-cm^{-1} region, but a strong band is seen near
500 cm^{-1}, which is attributed to the symmetric Si-O-Si stretching
mode. Silicones represent a case where the symmetry of the molecule
leads to a high degree of mutual exclusion between IR and Raman active
vibrations. The C-Si stretch is also observed in the Raman as a
medium intensity line at 710 cm^{-1}. The remaining polymers which contain aromatic groups consist of styrene systems, poly(phenylene oxide),
polysulfones, and epoxides.

5. Other Polymers

A large number of polymer types containing neither carbonyl nor
aromatic groups are found in the right-hand portion of the classification scheme in Figure 8. These can be subdivided with little difficulty according to particular functional groups which characterize
the species.

The presence of an OH or NH band near 3300 cm^{-1} gives a group which can be divided further by the presence of a strong band at 1670 cm^{-1}. This vibration is present in nylons, amino acids, and cellulosics, and the additional presence of a 1540-cm^{-1} amide band characterizes nylons and the amino resin. These vibrations are generally much stronger in IR than Raman spectra. (Although these polymer types contain a carbonyl group, they appear in this section because the amide C=O stretch is shifted out of the normal region due to mixing with N-H bending.) The cellulosics--nitrocellulose and cellophane--comprise the group not having the 1540-cm^{-1} band. The group which did not exhibit a 1670-cm^{-1} vibration contains some vinyl polymers, two more cellulosics, an amino acid, and an epoxide.

Of the polymers which do not exhibit OH or NH absorption, the initial classification is made by examining the spectrum for a strong 1430-cm^{-1} band. The members of this group which display a sharp 2220-cm^{-1} vibration (C≡N stretch) are acrylonitrile copolymers; and those which do not include vinyl, polyether, chloroprene, and epichlorohydrin polymers. The nitrile group also gives rise to a characteristic Raman band in the same region. This is a very good diagnostic peak in either spectrum, since the region is usually free from most other bands. The next group of the non-NH, OH-containing polymers is made by observation of a strong band near 1250 cm^{-1}. This arises from different functional groups: C-Si stretch in the silicones, C-O-C stretch in polyethers, C-F stretch in fluorinated polymers, and uncertain origin in polysulfides. Spectra of these polymers are sufficiently different in other regions to allow for identification by comparison with reference spectra. Useful bands in Raman spectra for characterization of silicones were mentioned earlier. The symmetric C-O-C stretch of polyethers produces a strong Raman line near 920 cm^{-1}. Raman spectra of polysulfides also provide useful information, the C-S and S-S stretches producing very strong lines at 650 and 510 cm^{-1}, respectively.

Those polymers which do not exhibit the sharp 1250-cm^{-1} band are further grouped by examination of the spectrum for a strong band in the 1000- to 1250-cm^{-1} range. Polymers which display this band include

fluorinated polymers (C-F stretch), poly(vinyl ethyl ether) (C-O-C
stretch), modified polypropylene and polysulfides (C-O-C stretch).
Raman spectra may be useful as already mentioned. The remaining
polymers, which are mainly hydrocarbon systems, can be divided into
two groups by the presence or absence of a sharp 1370-cm^{-1} band.
This band is observed when CH_3 groups are present and thus polypropylene, polyisoprenes, butyl rubbers, and a few others are grouped
on one side, and polyethylenes and butadiene and some of its copolymers are grouped on the other side.

      6. Conclusions

The application of the classification scheme presented in Fig.
8 is not completely clear-cut or foolproof, but it does simplify the
problem of identifying an unknown sample. Final identification must
always rest with matching the unknown spectrum with that of a reference standard. The process may be complicated by the presence of
plasticizers or other additives. Reference spectra of these compounds
are also available, however, and compensation can be made if the components are identifiable. The pyrolysis technique should also be
considered if difficulty is encountered by conventional means. With
the additional information which can now be obtained from Raman spectra, this technique should be utilized whenever instrumentation is
available.

B. Assignment of Normal Vibrations

This application of vibrational spectroscopy is one for which
it is nearly essential to have both the IR and Raman spectra of a
sample. Selection rules dictate that the IR and Raman spectra of a
compound may vary from total coincidence of all bands through various
degrees of partial coincidence to the case of mutual exclusion for
which there are no coincidences between IR and Raman spectra. The
results depend on the symmetry of the molecule. Some work has been
done using IR spectra only, because of the difficulty in obtaining
Raman data until recently, but this was done more as a necessity than
by choice. The availability of Raman information on polymers has
increased the application of vibrational studies to polymers.

Polymers are governed by the same rules as simple molecules, although sometimes the application of the rules is more difficult. As
is the case with simple molecules, all workers do not always agree on
assignments of normal vibrations nor, in fact, even observe all the
same lines in their spectra. Discussions giving two or three authors'
views on how the vibrational spectrum of a polymer should be assigned
are easily found in the literature.

The application of normal coordinate analysis to polymers has also
been accomplished in a few cases, although this type of work requires
special adaptation of existing theory for polymers. The classical
Wilson GF-matrix method [37] was extended to helical molecules by
Higgs in 1953 [38]. His work allowed for the determination of selection rules, intensities, and approximate frequencies in such molecules, his particular target being proteins. Other workers subsequently extended or modified his treatment to include other types of
structure [39,40].

Usually, it is necessary to know a considerable amount about the
structure of a polymer molecule before an assignment of normal vibrations can be made. Symmetry of the molecule, including the chemical
bonding structure and the arrangement of the atoms in space, is known
for the simpler polymer species. X-ray diffraction studies provide
much of this information. When the symmetry group of the molecule has
been determined, the information obtained from vibrational spectra can
be applied to the task of assigning all the vibrational species. Such
information as activity of a vibration, band intensity, IR dichroism,
and Raman depolarization values, in addition to group frequency correlations, makes the assignment of normal vibrations possible. Many
polymers exist in more than one configuration and/or conformation,
and different assignments are required for the different forms.

The simplest polymer systems have received the most attention.
Polyethylene has, naturally, been studied most extensively [41-46].
Other typical systems which have been treated include polypropylene
[47-50], poly(oxymethylene) [51-53], poly(vinylidene fluoride) [54],
and poly(tetrafluoroethylene) [55-58]. The two forms of polyglycine,

the simplest poly(amino acid), were also recently analyzed [59-61] with the intent of extending such treatment to more complex amino acids. Use has also been made of deuterated species whenever possible to assist in making assignments.

Polyethylene represents one of the extreme cases of structural symmetry. Its planar zigzag molecular chain has the high symmetry D_{2h}. Since it possesses a center of symmetry, there is mutual exclusion between the IR and Raman spectra. Polyethylene has long been of interest of spectroscopists, and early assignments of the vibrational spectra were based on IR data alone [41]. Raman data became available in 1957 [42], and increasingly more sophisticated treatments of the vibrational spectrum of polyethylene have appeared as technology has improved. Work on deuterated polyethylene aided in the understanding of the fundamental vibrations [44], but the controversy concerning the assignment of the B_{2g} methylene wag indicates the difficulties that may arise in the assignment of polymer spectra. The Raman line at 1415 cm^{-1} had originally been ascribed to this vibration [41], and this assignment was repeated by subsequent workers; but debate has continued through 1970 regarding the validity of this assignment. A weaker feature in the 1370-cm^{-1} region was later proposed as the B_{2g} wag [45], with the 1414-cm^{-1} line being a combination tone whose intensity is increased by Fermi resonance with the nearby fundamental. Examination of this region of the Raman spectrum under varying conditions provided additional information on this problem, but not all of it led to the same conclusion. Hendra [62] in 1968 found no features in the 1370-cm^{-1} region using laser excitation, and he supported the assignment of the 1417-cm^{-1} line as the B_{2g} wag. Snyder [63] observed the 1418-cm^{-1} line to be dependent upon crystal structure and, thus, not a fundamental. He said that Hendra's assignment was based upon symmetry of the isolated polyethylene chain rather than symmetry of the crystal and chose the 1370-cm^{-1} line as the B_{2g} wag. Carter [64] examined polarized Raman spectra of oriented polyethylene fibers and obtained three unique spectra by varying the polarization of the incident light and the analyzer. This analysis allowed him to determine the polarizability components of each band, and he found that the 1418-cm^{-1} line must be of the A_g symmetry species and the one at 1370 cm^{-1} probably either B_{2g} or B_{3g}. An examination of the Raman spectrum as a function of temperature led Boerio and Koenig [65] to the conclusion that the 1415-cm^{-1} line actually results from crystal field splitting of the A_{1g} CH_2 bending mode. The line was absent from the melt, and increased splitting occurred as the temperature was decreased.

This example of the simplest polymer material illustrates the complications that arise in assigning polymer spectra. Many more interactions need to be considered in determination of symmetry than for simpler molecules.

The recent treatments of vibrational spectra of polyethylene include extensions of the theory to cases of higher disorder. Snyder [46] considered the difficulties which arise in the treatment of a disordered chain as opposed to the regular one usually assumed. He used rotamers of liquid n-paraffins as models in his work and correlated normal coordinate calculations of the molecules to polyethylene. Zerbi et al. [66] considered the dynamics of a polymer chain containing a random distribution of conformational defects and proposed a method for setting up a dynamical matrix. The two models led to somewhat different conclusions and point to the importance of considering all possible aspects of polymer structure in normal coordinate treatments.

The already complex situation for polyethylene rapidly becomes more complicated as polymers with more substituent groups are considered. The addition of pendant methyl groups to the methylene chain presents the possibility of different rotational isomers in polypropylene. Helical syndiotactic polypropylene has an internal rotation configuration of trans, trans, gauche, gauche, denoted TTGG, with four chemical units to a repeat unit [47]. The corresponding isotactic polypropylene is arranged in alternate sequences of T and G and contains three chemical units per repeat unit [48]. This leads to different symmetry groups for the two forms and different numbers of normal modes as well. The syndiotactic form has the symmetry D_2 (104 normal vibrations), and isotactic polypropylene belongs to symmetry group C_3 (77 normal modes). In addition to these two forms having regular arrangements, atactic polypropylene--for which there is no fixed pattern of arrangement of methyl groups--is also possible. Crystalline modifications also exist for the various forms. For example, a crystalline form of syndiotactic polypropylene is less stable than the helical conformation and has a zigzag chain with repeat units arranged in a TTTT fashion. All of these differences may cause important changes in vibrational spectra which aid in determination of configuration and conformation, as will be discussed later. It is apparent here, however, that analysis of polymer vibrations is quite complex.

The various forms of polypropylene have been studied and assignments have been made using IR and Raman data. The 77 normal modes (25 A + 26 E) of helical isotactic polypropylene are all active in both the IR and Raman spectra. Tadokoro et al. [48] made assignments for the IR spectrum and performed normal coordinate calculations. Only limited Raman data were available at the time, but the laser Raman spectrum of polypropylene was subsequently reported and found to be in good agreement with Tadokoro's IR results [50]. Koenig and Vasko [49] reexamined the spectra of isotactic polypropylene using both IR and Raman data. For syndiotactic polypropylene, the 104 normal vibrations are made up of 104 Raman-active and 78 IR-active modes. Assignment of the IR spectrum has been made [47].

Another polymer which has received considerable attention is poly(oxymethylene). A study of the laser Raman spectrum of hexagonal (D_9) and orthorhombic (D_2) crystalline forms by Zerbi and Hendra [51] gave support to spectral assignments which had been made previously based on IR results alone. They made use of the deuterated compound for comparison, as did Sugeta et al. [52]. These latter workers reported some discrepancies between their observed spectra and those of Zerbi and Hendra. Subsequent work by Hendra et al. [53] compared assignments of poly(oxymethylene) with spectral data observed for poly(thiomethylene) and poly(selenomethylene).

Poly(tetrafluoroethylene) is also a very simple polymer whose vibrational spectrum has been analyzed. There are two helical conformations of the crystalline form of this polymer, but the symmetry of both is low and the same IR and Raman activity of normal vibrations is predicted. Koenig et al. [55-57] made assignments and a normal coordinate analysis, and Peacock et al. [58] reexamined the spectra from a standpoint of effects of varying crystallinity over a temperature range of -150 to 60°C. They used these results to recommend revisions in some of the earlier assignments. A comparison of the analysis of the spectrum of PTFE with that of polyethylene shows the interesting effects which arise from conformational differences in very chemically similar systems.

A somewhat more complex material for vibrational analysis is polyglycine, but it is of particular interest because of its relation to polypeptide analogs. Two conformational forms (I--extended planar structure [60] and II--helical conformation [59,61]) have been studied. Hydrogen bonding also plays a part in the assignment of vibrational frequencies in these molecules because N-H • • • O and C-H • • • O links are found between neighboring chains.

Thus, in spite of the complexities presented by interpretation of vibrational spectra of polymers, there has been much interest in this kind of application. Increasingly more sophisticated treatments of polymer vibrations have appeared as the theory has been extended to cover the peculiarities unique to polymers. The capability of obtaining good Raman data has, of course, played a primary part in the extension of vibrational analyses. It appears that as the theory of polymer spectra continues to advance workers will hasten to apply the results to more and more complex systems.

C. Quantitative Analysis of Polymer Systems

The modification of properties of polymers by copolymerization is very common in polymer technology. Polymers have many important applications, and the capability of imparting desired properties has

played a large part in creating this widespread usage. A polymer
which is selected for a particular application may often be made
nearly perfect for that application by copolymerizing one or more
other compounds with it. This technique virtually gives an endless
number of polymer systems that may be prepared. This has led to the
need for a technique for quantitative analysis of components in copolymer systems. IR spectroscopy is quite well suited to this application, and numerous studies have been performed in this regard.

Another use which has been made of the quantitative application
of IR is the determination of end groups in polymers. The basis of
this and other quantitative applications is measurement of bands
which are unambiguously assigned to a particular functional group.

The quantitative use of IR requires application of the Beer-
Lambert law

$$\ln I_o/I = abc$$

where the absorbance ($\ln I_o/I$) is directly proportional to the concentration (c) of the absorbing species. If the absorption coefficient (a) is known for a vibration and the sample thickness (b) can
be determined, concentration can be calculated. However, the Beer-
Lambert law is not always obeyed and the more general application
in polymer systems involves making plots of band absorbance ratios
with composition for standard samples. The value of the absorption
coefficient is greatly affected by environmental changes. This has
quite an effect in copolymers, because conditions of polymerization
determine length of copolymer segments, orientation of pendant groups,
branching, etc., all of which in turn affect the environment for the
functional group and thus the absorption coefficient.

Care must be exercised in the quantitative application of IR
spectroscopy to polymers to insure that samples of similar derivation
are compared for greatest accuracy. Different methods of preparing
standard samples for calibration purposes have been used.

Raman spectroscopy also offers a quantitative capability, but
the application of this aspect of Raman scatter has not received much
attention in polymeric systems. A different principle is utilized,

and measurements are comparative rather than absolute. Reproduction
of exact instrument settings from one sample to another is of primary
importance. Because of the different sensitivities of the IR and
Raman techniques to different structural characteristics, the quantitative application of Raman spectroscopy could prove quite useful in
the study of some systems where IR cannot provide bands which are
sufficiently widely separated or for determining relative amounts of
specific chemical groups, i.e., nonpolar or skeletal structure. For
example, Koenig [67] has shown that polymerization of butadiene proceeds by several competing mechanisms. The polymer products all have
unsaturation and the C=C stretching band can be more easily followed
by Raman than by IR since Raman is much more sensitive (in fact, the
C=C stretch in the trans-1,4 product is IR-inactive). For copolymers
it is possible to determine the relative amounts of monomers and
their distribution. In the case of vinyl chloride/vinylidene chloride
copolymers [68], it has been possible to determine quantitatively the
percent vinyl chloride in the copolymer and the relative amounts of
VC-VC, VC-VC-VC, VDC-VDC, VDC-VDC-VDC, and VDC-VDC-VDC-VDC (where
VC = vinyl chloride and VDC = vinylidene chloride) linkages in the
polymer.

Application has been made of IR spectroscopy to both absolute
and comparative quantitative studies. A recent study [69] of the
analysis of acrylic polymers modified by copolymerization made an
absolute determination of the absorption coefficient (absorptivity)
of the 700-cm^{-1} styrene band and then used this value in determining
styrene content in styrenated acrylic polymers. In this method,
styrene solutions of known concentrations were used for measurement
of absorbance values and determination of the absorptivity was made
by plotting absorbance against sample concentration. These workers
reported the application of this absorptivity value to acrylic systems
containing copolymerized styrene to be better than 95% accurate for
styrene contents in the 5 to 70% range. The advantage of this type
of quantitative application is that any future workers could evaluate
styrene concentration in any system by making one measurement, the

absorbance of a prepared solution of the polymer in a cell of known
pathlength. Criticism of such a transfer of information from one
system to another can be raised, however, because of the dependence
of the absorptivity value on surroundings of the functional group.
In this particular application, the environmental effect may be
minimal since the band being used for the determination is an out-
of-plane ring-bending mode. The phenyl ring will be a pendant group
in any copolymer system, and its nearest neighbor will not be affected by the polymerization. Environmental changes will be at least
one atom removed.

An example of a study performed taking into account more of the
variable factors affecting surroundings is a determination of composition in ethylene-propylene copolymers using ^{14}C-labeled species
[70]. In this work, polymers were prepared starting with either
labeled ethylene or propylene. In this way, measurements of concentration or IR absorption were made for groups having exactly the
same environment. Copolymers of 26 different compositions were prepared, some of which contained labeled ethylene and some labeled
propylene. In this type of study a plot of a ratio of band absorbances vs. concentration (in this case determined by radioactive
counting) is prepared as a calibration curve for use by comparison
with any future unknown samples.

Another application of quantitative IR spectroscopy was made by
Tompa [71] in the determination of content of terminal functional
groups in polybutadienes. Samples of various origins containing
either carboxyl or hydroxyl terminal groups were titrated to determine the concentration of the functional groups. IR measurements
then allowed the absorption coefficient to be determined, and calibration plots were made for use in evaluating functional group content in future samples. Several interesting points had to be considered in this work. One was the interference with hydroxyl band
measurement caused by a phenolic antioxidant. Since the composition
of the additive and its concentration were known, compensation could
be made by measurement of its absorptivity. Another feature of the

study was that several classes of hydroxyl-terminated polybutadienes were observed. This required that calibration curves be prepared for each class; it illustrates the differences that can be observed to be caused by surroundings. Hydrogen bonding was another factor which had to be considered. Graphical integration of bands had to be used if hydroxyl groups displayed any appreciable hydrogen-bonding species, but peak height measurements could be used if there was none.

One further illustration of the use of IR spectroscopy as a quantitative tool is some work that was done using near-IR measurements. An application was made to physical blends of polyethylene and polypropylene homopolymers by measuring absorbance ratios of combination and overtone bands rather than fundamentals [72]. This use presents the possibility of obtaining readings in a region of the spectrum where there may be less overlapping of bands arising from different components. Four ethylene-propylene rubber copolymers were also measured, and it was found that the calibration curve was different for these systems than for the physical blends. This again points out the effect of environment on absorptivity. An indication of the differences caused by the use of high-density polyethylene instead of low-density material in the blends was also given.

The use of normal IR spectroscopy for analysis of fibers, coatings, etc., presents obvious problems with transmission of radiation through the material. These problems can be overcome by the use of ATR. Wilks [73] and Wilks and Cassels [74] have reported the applicatio of ATR to quantitative problems and to analysis of cotton-nylon mixtures in particular. A recent publication by Basch and Tepper [75] demonstrates the use of ATR to the quantitative analysis of polyester-wool mixtures. In this case, standard curves of absorbance ratios vs. percent polyester were constructed using known mixtures. The curves deviated markedly from Beer's law plots. The 1714-cm^{-1} band of the polyester and the 1520-cm^{-1} wool band were used in the analysis. Test samples deviated from the known composition by only 2 to 4%.

Some points should be discussed which are common to most quantitative applications of IR spectroscopy. Since all of these applications depend upon measurement of the absorbance of an IR band, the
method by which this measurement is made is important. One of two
methods is usually used: measurement of the peak height or of the
area covered by a band. The precise measurement required is the
absorbance (ln I_o/I) at a monochromatic wavelength. Slit openings
normally employed in spectrophotometers prevent the light from actually being monochromatic, however, and the methods of band area and
peak height measurement are approximations. Use of the band area
takes into account the spreading of the absorption over a range of
frequencies. As mentioned earlier, this method also is quite useful
when vibrations affected by hydrogen bonding are being measured.
The peak height method is satisfactory for most applications in
polymers because the slit width is much smaller than the band width.

In either of these cases of measuring the absorbance, establishing the base line from which the measurement is to be made is another
important consideration. The ideal case would be that all bands be-
gin from a base line of 100% T. This is never the actual case,
however, because of the normal scattering of radiation caused by
samples. In addition, bands of interest are not always completely
isolated from other absorptions and the base line for the band may
be shifted by the interference. Different methods of drawing base
lines are used, but the main point of consideration should be con-
sistency among samples used in the same determination [9].

Some mention has already been made of methods of preparing
standards for calibration purposes and the hazards involved in trans-
ferring information between systems that are not identical. Another
consideration in this regard is the choice of band to be used for
measurements. Aside from the desire to have the vibration as far
removed as possible from interfering vibrations, the origin of the
vibration itself needs to be considered. Stretching vibrations tend
to be less sensitive to nearby atoms than do bending modes and,

therefore, more reliable when transfer of information is being made. It is not always possible to choose samples and bands which are ideal from all standpoints, but the closer these conditions can be approximated, the more accurate the information obtained will be.

D. Study of Chemical and Physical Changes

An important application of vibrational spectroscopy has been in the study of chemical and physical changes occurring in polymeric systems. These properties are of great importance in the use of polymers, and information concerning changes in them assists in the commercial applications of polymers. The wide range of uses of polymers has caused quite a variety of applications of vibrational spectroscopy to have been made to the study of changes in properties. Physical effects, such as applying stress, rolling, and abrading, and chemical changes such as vulcanization, pyrolysis, and photodegradation, all lead to detectable changes in molecular vibrations; consequently, IR and Raman spectra provide a means of monitoring the effects caused.

This is another application of vibrational spectroscopy for which Raman has received far less attention than IR, again mainly for the same reasons that hindered its application to polymers in general. Some recent applications using the new Raman capability have indicated the value that this technique may offer.

Any action which results in a change in bond length, bonded groups, or even orientation of groups or molecules will cause some change in molecular vibrations. A great many of these effects have been studied using vibrational spectroscopy. It is not appropriate to discuss all of them here, but some examples will serve to illustrate the correlations which can be obtained.

One area of study has been that of trying to determine what takes place in the actual preparation of polymers or their systems, either the mechanism of the reaction or the types of bonds formed in it. Polymerization itself provides an area of interest for study. An example of this is the study of styrene polymerization by Raman

spectra [76]. Observation of a decrease in the intensity of the
1636-cm^{-1} C=C stretch, compared with the 622-cm^{-1} band as an internal
standard, is a method of determining the kinetics of the polymerization
reaction. Information concerning the preparation of polymethylene from
carbon monoxide and hydrogen [77] on ruthenium catalysts was gained
from IR spectra of products isolated in the reaction. Metallic ruthenium gave high-molecular weight hydrocarbon compounds, while ruthenium
carbonyls were formed when unreduced RuO_2 was used. Vulcanization
of rubbers has been the subject of recent Raman [78,79] and IR [80,
81] studies. As has been mentioned in an earlier section, the Raman
effect is particularly sensitive to vibrations involving S-S and C-S
groups. This fact was used to analyze the types of linkages formed
by sulfur vulcanization of cis-1,4-polybutadiene. Tentative assignments of a number of lines were made for vibrations corresponding to
disulfide, polysulfide, and five- and six-membered thioalkene and
thioalkane structures [78]. In addition, the structures found in the
vulcanizate were found to vary with the length of time and temperature
of the vulcanization. Information was also obtained from the C=C
double-bond stretching region regarding main chain modifications [79].
Here correlations were made for cis-1,4 (1650 cm^{-1}), trans-1,4 (1664
cm^{-1}), vinyl (1640 cm^{-1}), and conjugated triene (1623 cm^{-1}) groups.
Net trans concentration was found to increase upon vulcanization.
An IR study [80] of pressure vulcanization of butadiene-styrene rubber also provided information regarding the course of vulcanization.
Intensity of out-of-plane C-H bending modes indicated that trans-1,4
addition decreased while 1,2 addition remained constant. All of this
information provides an insight to the isomerization and cross-linking
which takes place during vulcanization.

     Reactions which take place in polymeric materials as a result of
applied physical forces have provided another area of interest. Many
of the uses of polymeric materials involve exposure to various physical effects. Thus, there has been considerable interest in methods
of studying changes that may be caused by such exposure which might
decrease the stability of the system. Many of the changes are

accompanied by oxidation when the exposure is carried out in air.
ATR can be particularly useful in these cases, since such oxidation
will usually be concentrated at the surface of the material. Aging
and weathering effects on polymers represent an area in which vibrational information can be correlated with observed changes in properties to give some understanding of the factors causing the changes
and possible means of correcting for them. Examples of this application are a weathering study of polyethylene exposure to sea air [82]
and the aging behavior of a high-temperature poly(arylsulfone) resin
containing no aliphatic linkages [83]. In the latter study, IR spectra of samples aged at 275 and 300°C on aluminum bars showed the
formation of polymeric hydroxyl groups, cross-linking (change from
disubstituted to trisubstituted rings), and acid and/or quinone carbonyl stretches. Irradiation of polymers has also provided an area
of interest for researchers. ^{60}Co γ-rays produced IR changes indicating formation of C=O groups and trans CH=CH structure in polyethylene [84,85]. Irradiation of poly(methyl methacrylate) with a
mercury discharge lamp produced photodegradative effects that showed
several changes in the IR spectrum. A decrease was observed for absorptions of ester, methyl, and methylene groups and an increase for
MeOAc; no change in the C=C absorption was noted. Based on these observations, a mechanism for photodegradation was proposed [86]. The
utility of ATR was demonstrated in a study of irradiation of polyethylene [87] where a C=O absorption at 1710 cm^{-1} was found to increase linearly with irradiation time. Applied stress is an influence
which results in changes in the IR spectrum which have more physical
correlations [88-91]. Spectra of samples of polyethylene and polypropylene were determined before and after applying loads for use
in measuring the kinetics of the mechanical degradation of polymers.
Bands in polyethylene were observed to increase in intensity under
loads of 50 kg/mm^2 applied for 1 min [91]. Another study of stress
on chemical bonds showed that shifting of frequency of skeletal vibrations occurred under stress [89].

Chemical reactions of polymers have also been of interest. Often information obtained from vibrational spectra has aided in the determination of reaction mechanisms [92-94]. For example, IR spectra of polyisobutylene during radical chlorination showed changes affecting CH_2 and CH groups as the amount of combined chlorine increased, in addition to some degradation accompanying the chlorination process, as indicated by an increase in low-molecular weight polymer content [94]. It was also determined that the route of the reaction did not seem to be dependent upon the kind of initiator. Completion of a reaction can also be followed using spectra, as was done in a hydrogenation study on butadiene-styrene copolymers [95]. Decrease in the ratio of absorbance of two IR bands indicated the reduction of unsaturation in the polymer.

Hydrogen bonding is another case of interest that has been studied by vibrational spectroscopy. IR bands sensitive to hydrogen bonding between NH and C=O groups in Kapron studied as a function of temperature showed decreased absorbance and increased half-width with increasing temperature [96]. Shifting of some bands to different frequencies also occurred. Intramolecular hydrogen bonds between COOH groups in poly(methacrylic acid) were observed to increase when methanol was added to solutions of the polymer [97]. The IR study of solutions in methanol, water, and their mixtures indicated that no hydrogen bonds were formed with methanol or water. Specificity of hydrogen bonding in nucleic acid purines and pyrimidines has also been observed [98]. An IR study of the association of hypoxanthine with derivatives of guanine, cytosine, adenine, thymine, and uracil in chloroform solution showed that strong association occurred in the case of cytosine, weak association with adenine, and only very slight association with any of the others.

These examples of recent work, while not illustrating every type of application of vibrational spectroscopy that has been made for physical and chemical changes, do give an indication of the utility that is available to the researcher. The possibilities of

application in actual systems are as varied as the imagination allows. The special capabilities afforded by ATR should also be kept
in mind for extension of data to special systems.

E. Study of Conformation
   in Polymers

Some mention has already been made of the effect of polymer
configuration and conformation on the vibrational spectrum in the
discussion of assignment of spectra. This effect has also led to a
very important application of vibrational spectroscopy in the analysis of conformational effects. Quite different physical properties
are often associated with the same polymer in different conformations.
An example is the solubility of atactic polystyrene in chloroform as
compared with the insolubility of isotactic polystyrene in the same
solvent [99]. X-ray diffraction probably provides the most precise
information regarding structure in polymers, but vibrational spectroscopy is also important in this application, particularly for
systems that do not lend themselves to x-ray examination or for cases
in which the ordering in the molecule is not over a sufficiently
large range to be detectable by x-ray methods, since the techniques
and sensitivities differ for the two methods [100].

Differences in configuration of single repeat units in a polymer
lead to rotational isomerism. Isotactic polymers are formed when
polymerization occurs with monomer units joining so that substituents
are positioned on the same side of the chain. Syndiotactic polymers,
on the other hand, have the substituent groups regularly alternating
on opposite sides of the backbone. Atactic polymerization represents
the case for which there is no pattern of arrangement of side groups.
Arrangement of sequence of repeat units results is a second kind of
ordering. Different conformations are possible for polymers depending upon the constraints placed upon the molecule by the surroundings.
Crystallization is one of the main factors affecting the conformation
a polymer assumes, but other factors such as temperature, hydration,
hydrogen bonding, and catalysis also play a part. Most often, for

polymers which exist in more than one crystalline modification, the
method of preparation determines the conformation. Polybutene-1,
however, is an example of a polymer which undergoes crystal-crystal
transformations between its three crystalline modifications [101].

These properties of polymers have been of considerable interest
to vibrational spectroscopists and many studies have been performed
using both IR and Raman techniques. Some interesting correlations
have been made by studying a polymer at different temperatures or
by examining the same polymer prepared under varying conditions.
Spectral changes which occur usually involve intensity changes,
splitting of bands, and polarization differences between different
forms.

One polymer conformation that is quite common and of particular
interest is the helical form. Its presence can be observed in the
IR by the occurrence of bands split into two components having opposite polarization [102]. Also, if a polarized Raman line is coin-
cident with a parallel IR band and a depolarized Raman line with a
perpendicular IR band, a helical conformation is probably present
[67].

One application of vibrational analysis for studying conformation is determination of symmetry group by differences in characteristics of normal vibrations. Koenig [100] has given a good analysis
of the situation for monosubstituted vinyl polymers. The point
groups indicated for such polymers are as follows:

Syndiotactic, planar	C_{2v}
Syndiotactic, twofold helix	D_2
Syndiotactic, threefold helix	D_3
Syndiotactic, n-fold helix (n > 3)	D_n
Isotactic, planar	C_s
Isotactic, threefold helix	C_3
Isotactic, n-fold helix (n > 3)	C_n
Atactic	C_1

The differences in vibrational activity, Raman polarization, and IR
dichroism for these symmetry groups can lead to determination of

structure from thorough examination of IR and Raman data. A brief
summary of the spectra of some of the vinyl polymers relative to
their conformation is given by Koenig [100].

A general expression for the factor group of monosubstituted
vinyl helical molecules is $C_{2t\pi/u}$, where t represents the number of
turns in the helix in which u number of units reestablish the original position along the chain [100]. The shifting of frequency of
some modes with helix angle can be used to estimate the values of t
and u. In polybutene-1, form I was known to exist as a 3_1 (three-
fold) helix and form II as an 11_3 helix. The helical nature of
form III was uncertain, but an estimate that it approximated 10_3 was
made after an examination of the Raman spectrum and a normal coordinate calculation [101]. Certain normal modes were found to shift
as a function of helix angle, and a plot of frequency vs. angle using
known values for I and II allowed a prediction to be made for III.

Some polymers exhibit an unwinding of the helix as a result of
changes in temperature. Poly(tetrafluoroethylene) is such a polymer,
and its helical structure below 19°C contains 13 CF_2 groups in 6
turns and at temperatures above 19°C 15 groups in 7 turns of the
helix [103]. Such differences should produce effects that could be
observed in IR and Raman spectra, and this is the case. Although
the changes are not as dramatic as might be expected, significant
intensity changes are observed for a Raman pair and an IR pair of
bands. One band of the pair is seen to disappear from the spectrum
at low temperatures while the other increases in intensity. Normal
coordinate calculations performed for the two structures (as well
as for a planar zigzag structure) gave calculated frequencies showing
little difference between the two forms, consistent with experimental
results [104]. The planar zigzag form may possibly exist at high
temperatures and pressures.

An IR study of isomerism in polypropylene found that a band at
997 cm^{-1} was sensitive to isotactic helical content [105]. Because
the band was not present in the spectrum when the sequence of isotactic units was less than 10, it was concluded that helical conformation could not be assumed by fewer than 10 consecutive isotactic

monomer units. The IR frequency of C=O and C≡N stretching modes
has been shown to be dependent upon stereoregularity in poly(methyl
methacrylate) and poly(methacrylonitrile) [106]. Various polymers
of those mentioned above were prepared using different organometallic
catalysts to give different proportions of isotactic content. It
was observed that higher frequencies of the stretching modes corres-
ponded to higher isotacticity (determined by nmr).

A subject of special interest in conformational analysis is
biopolymers. Water is an important component of biological systems,
and its function in structural determination of proteins and nucleic
acids provides an interesting case for study by vibrational spectroscopy [107,108]. Raman spectroscopy is particularly well suited to
such studies. The sensitivities of IR and Raman spectra to conformational variations as applied in other polymers can be applied in the
same fashion in biopolymers.

Another correlation to be found in vibrational spectra is with
density [109-110]. The width of the carbonyl stretching band in the
Raman spectrum of poly(ethylene terephthalate) was found to decrease
with increasing density of the polymer. Because only the C=O vibration changed bandwidth as the amorphous content changed, a mechanism
of resonance stabilization was proposed [110]. If crystallization
promotes planarity, resonance between carbonyl groups and the phenyl
ring is possible:

and a narrower band results.

Many more studies involving polymers in the form of melts, solutions, stretched fibers, etc., provide interesting examples of the
effects that changes in polymer conformation have on molecular vibrations. This sensitivity of vibrational spectroscopy has been applied
widely and should continue to be extended as techniques for preparing
more polymers having precise configurations and conformations are
developed.

V. SUMMARY

IR and Raman techniques are extremely useful tools for identification, analysis, and structural determination of polymeric materials. Raman spectroscopy is rapidly becoming a popular technique due to the introduction of the laser as an exciting source; and fluorescence, which has been particularly troublesome for polymers and biological materials, may be reduced or eliminated by future developments. The advantages of Raman spectroscopy include (1) the ease of sample preparation, (2) the facility to identify and analyze important skeletal and nonpolar portions or changes in polymers, e.g., those resulting from vulcanization, and (3) the ability to study aqueous solutions of polymers. It is particularly useful for biological systems. IR spectra, on the other hand, are useful for studying changes in polar groups such as carbonyl formation during thermal or irradiative oxidation of polymers. ATR is a special technique which is useful in differentiating and characterizing polymer surfaces and coatings from bulk or substrate. It is also useful in obtaining spectra of materials which are not transparent and cannot be easily modified. Another special technique which is applicable to the analysis of uncooperative materials is the spectroscopic examination of pyrolysis products.

Many applications have been made of vibrational analyses to polymer systems, and many more remain to be made. The brief survey of typical applications given here with examples serves to illustrate the great variety of uses that can be made of IR and Raman spectra. Not all types of reported studies have been mentioned, but an attempt has been made to cover the most frequently used techniques. Identification of polymers has long been a popular use of IR spectroscopy, and Raman is now finding its place in this area as well. The classification scheme discussed in this chapter provides a systematic method of determining polymer composition from spectral information. Examples of assignment of normal vibrations in some simple polymer systems illustrate some of the accomplishments that have been made in this area and some of the complications that are encountered in

such studies. Relatively complex systems have been analyzed in detail by extension of basic theories to polymer cases. Quantitative
analysis of copolymers is an area where considerable use has been
made of IR spectroscopy. Raman spectroscopy is now also being used
in this regard, as is the special IR technique of ATR. Changes in
physical and chemical properties provide a great variety of oppor-
tunities for correlations with vibrational spectral properties.
Mechanisms of reactions, bonding changes associated with changes in
properties, and effects of hydration are representative types of
applications that have been made. Finally, the utility of vibrational information regarding conformation in polymers is an area where
work is rapidly expanding. Polymers present quite different problems
from simpler molecules, and vibrational spectra have been shown to
be quite sensitive to effects of configuration and conformation. A
special area of interest is biopolymers and the relation between
hydration and conformation.

Thus, the vast amount of work which has been performed using
IR and Raman spectroscopy shows the value they present in analyses
of polymers. Many new areas of application will also appear as uses
and modifications of polymers are extended. Raman spectroscopy especially offers advantages in some areas not previously amenable to
IR techniques. The advantages of sample handling and acceptability
of water as a solvent should prove particularly useful in polymer
applications.

REFERENCES

1. R. F. Schaufele, *J. Polymer Sci.*, *4D*, 67 (1970).

2. R. E. Hester, *Anal. Chem.*, *44*, 490R (1972).

3. H. J. Sloane, *Appl. Spectrosc.*, *25*, 430 (1971).

4. J. L. Koenig, *Appl. Spectrosc. Rev.*, *4*, 233 (1971).

5. J. P. Luongo, *Appl. Spectrosc.*, *25*, 76 (1971).

6. D. O. Hummel, *Infrared Analysis of Polymers, Resins, and Addi-
    tives: An Atlas, Vol. 1: Plastics, Elastomers, Fibers, and
    Resins, Part 1*, Wiley-Interscience, New York, 1971, 224 pp.

7. Infrared Spectroscopy Committee of the Chicago Society for Paint Technology, *Infrared Spectroscopy, Its Use in the Coatings Industry,* published by the Federation of Societies for Paint Technology, Phila., Pa., 1968.

8. R. F. Schaufele, *Trans. New York Acad. Sci., 30,* 69 (1967).

9. W. J. Potts, *Chemical Infrared Spectroscopy, Vol. 1 - Techniques,* John Wiley and Sons, Inc., New York, 1963, 322 pp.

10. D. O. Hummel, *Infrared Spectra of Polymers in the Medium and Long Wavelength Regions,* Wiley-Interscience, New York, 1966, 207 pp.

11. P. A. Wilks, Jr. and T. Hirschfeld, *Appl. Spectrosc. Rev., 1,* 99 (1967).

12. W. R. Feairheller, Jr. and W. J. Crawford, *Develop. Appl. Spectrosc., 7B,* 43 (1968).

13. V. P. Panov and V. I. Zheleznov, *Zavod. Lab., 36,* 1535 (1970).

14. R. A. Burley and G. R. Woollerton, *Can. Spectrosc., 16,* 67 (1971).

15. B. Bott and F. Vernon, *Spectrochim. Acta, 26A,* 1633 (1970).

16. J. R. Nielsen, *J. Polym. Sci., 7C,* 19 (1964).

17. Block Engineering, private communication.

18. J. L. Koenig, M. M. Coleman, J. R. Shelton, and P. H. Starmer, *Rubber Chem. Technol., 44,* 71 (1971).

19. J. P. Luongo, *J. Polym. Sci., 42,* 139 (1960).

20. H. J. Sloane, *Polymer Characterization: Interdisciplinary Approaches,* (C. Craver, ed.), Plenum Press, New York, 1971, pp. 15-36.

21. R. E. Kagarise and L. A. Weinberger, *Infrared Spectra of Plastics and Resins,* NRL Report 4369, May 26, 1954.

22. S. S. Stimler and R. E. Kagarise, *Infrared Spectra of Plastics and Resins. Part 2--Materials Developed Since 1954,* NRL Report 6392, May 23, 1966.

23. D. S. Cain and S. S. Stimler, *Infrared Spectra of Plastics and Resins. Part 3--Related Polymeric Materials (Elastomers),* NRL Report 6503, Feb. 28, 1967.

24. R. A. Nyquist, *Infrared Spectra of Plastics and Resins,* Dow Chemical Co., Midland, Mich., May 3, 1961.

25. W. H. T. Davison and G. R. Bates, *Rubber Chem. Technol., 30,* 771 (1957).

26. D. Hummel, *Rubber Chem. Technol., 32,* 854 (1959).

27. J. Haslam and H. A. Willis, *Identification and Analysis of Plastics,* van Nostrand, Princeton, 1965, pp. 373-469.

28. L. J. Bellamy, *Infrared Spectra of Complex Molecules,* John Wiley and Sons, Inc., New York, 1958, 425 pp.

29. D. S. Cain and A. B. Harvey, *Raman Spectroscopy of Polymeric Materials. Part 1--Selected Commercial Polymers,* NRL Report 6792, Dec. 26, 1968; also in *Develop. Appl. Spectrosc., 7B,* 94 (1968).

30. T. Kajiura and S. Muraishi, *Nippon Kagaku Zasshi, 89,* 1187 (1968).

31. G. J. Clark and R. Granquist, *Rubber Age, 1968,* 77.

32. D. A. MacKillop, *Anal. Chem., 40,* 607 (1968).

33. D. L. Harms, *Anal. Chem., 25,* 1140 (1953).

34. J. W. Cassels, Identification of Coatings or Treatments on Glass Fibers by Pyrolysis-Infrared Spectroscopy, presented at 19th Southeastern Regional Meeting, ACS, Nov. 1, 1967.

35. J. W. Cassels, Infrared-Pyrolysis Studies of Synthetic Fibers, presented at 18th Pittsburgh Conf. on Anal. Chem. and Appl. Spectrosc., March 1967.

36. Two such units are produced by Barnes Engineering Co., Stamford, Conn., and Wilks Scientific Corp., South Norwalk, Conn.

37. E. B. Wilson, Jr., J. C. Decius, and P. C. Cross, *Molecular Vibrations,* McGraw-Hill, New York, 1955, 388 pp.

38. P. W. Higgs, *Proc. Roy. Soc., A220,* 472 (1953).

39. H. Tadokoro, *J. Chem. Phys., 33,* 1558 (1960).

40. T. Miyazawa, J. Ideguchi, and K. Fukushima, *J. Chem. Phys., 38,* 2709 (1963).

41. S. Krimm, C. Y. Liang, and G. B. B. M. Sutherland, *J. Chem. Phys., 25,* 549 (1956).

42. J. R. Nielsen and A. H. Woollett, *J. Chem. Phys., 26,* 1391 (1957).

43. J. R. Nielsen, *J. Polym. Sci., 7C,* 19 (1964).

44. R. G. Brown, *J. Chem. Phys., 38,* 221 (1963).

45. R. G. Snyder, *J. Mol. Spectrosc., 23,* 224 (1967).

46. R. G. Snyder, *J. Chem. Phys., 47,* 1316 (1967).

47. T. Miyazawa, *J. Chem. Phys., 43,* 4030 (1965).

48. H. Tadokoro, M. Kobayashi, M. Ukita, K. Yasufuku, S. Muraishi, and T. Torii, *J. Chem. Phys., 42,* 1432 (1965).

49. J. L. Koenig and P. D. Vasko, *Macromolecules, 3,* 597 (1970).

50. R. F. Schaufele, *J. Opt. Soc. Amer., 57,* 105 (1967).

51. G. Zerbi and P. J. Hendra, *J. Mol. Spectrosc., 27,* 17 (1968).

52. H. Sugeta, T. Miyazawa, and T. Kajiura, *J. Polym. Sci., 7B,* 251 (1969).

53. P. J. Hendra, D. S. Watson, and M. Mammi, *Spectrochim. Acta,* *28A,* 351 (1972).

54. F. J. Boerio and J. L. Koenig, *J. Polym. Sci.,* *9A-2,* 1517 (1971).

55. J. L. Koenig and F. J. Boerio, *J. Chem. Phys.,* *50,* 2823 (1969).

56. M. J. Hannon, F. J. Boerio, and J. L. Koenig, *J. Chem. Phys.,* 50, 2829 (1969).

57. F. J. Boerio and J. L. Koenig, *Bull. Amer. Phys. Soc.,* *14,* 406 (1969).

58. C. J. Peacock, P. J. Hendra, H. A. Willis, and M. E. A. Cudby, *J. Chem. Soc.,* *1970A,* 2943.

59. R. D. Singh and V. D. Gupta, *Spectrochim. Acta, 27A,* 385 (1971).

60. V. D. Gupta, S. Trevino, and H. Boutin, *J. Chem. Phys.,* *48,* 3008 (1968).

61. M. Smith, A. G. Walton, and J. L. Koenig, *Biopolymers, 8,* 29 (1969).

62. P. J. Hendra, *J. Mol. Spectrosc.,* *28,* 118 (1968).

63. R. G. Snyder, *J. Mol. Spectrosc.,* *31,* 464 (1969).

64. V. B. Carter, *J. Mol. Spectrosc.,* *34,* 356 (1970).

65. F. J. Boerio and J. L. Koenig, *J. Chem. Phys.,* *52,* 3425 (1970).

66. G. Zerbi, L. Piseri, and F. Cabassi, *Mol. Phys.,* *22,* 241 (1971).

67. J. L. Koenig, *Chem. Technol.,* 411 (1972).

68. M. Meeks and J. L. Koenig, *J. Polym. Sci.,* *9,* 717 (1971).

69. D. G. Anderson, K. E. Isakson, D. L. Snow, D. J. Tessari, and J. T. Vandeberg, *Anal. Chem.,* *43,* 894 (1971).

70. C. Tosi, M. P. Lachi, and A. Pinto, *Makromol. Chem.,* *120,* 225 (1968).

71. A. S. Tompa, *Anal. Chem.,* *44,* 628 (1972).

72. T. Takeuchi, S. Tsuge, and Y. Sugimura, *Anal. Chem.,* *41,* 184 (1969).

73. P. A. Wilks, Jr., *Appl. Spectrosc.,* *23,* 63 (1969).

74. P. A. Wilks, Jr. and J. W. Cassels, *Develop. Appl. Spectrosc.,* *7B,* 280 (1968).

75. A. Basch and E. Tepper, *Appl. Spectrosc.,* *27,* 268 (1973).

76. A. L. Bortnichuk and V. K. Shtyrkov, *Khim. Kinet. Termodin. Reakts. Krekinga Polim.,* *1968,* 70.

77. H. Pichler, H. Meier, W. Gabler, R. Gaertner, and D. Koussis, *Brennstoff.-Chem.,* *48,* 266 (1967).

78. J. L. Koenig, M. M. Coleman, J. R. Shelton, and P. H. Starmer, *Rubber Chem. Technol.,* *44,* 71 (1971).

79. J. R. Shelton, J. L. Koenig, and M. M. Coleman, *Rubber Chem. Technol.*, *44*, 904 (1971).

80. L. P. Semenova, N. A. Klauzen, M. S. Fel'dshtein, and V. A. Zhukova, *Zh. Prikl. Specktrosk.*, *6*, 759 (1967).

81. L. A. Murashova, A. A. Chekanova, V. G. Epshtein, and Yu. A. Yakovlevskaya, *Vysokomolekul Soedin*, *10B*, 750 (1968).

82. T. Simionescu, E. Sinchievici, and E. Pirligras, *Materie Plastiche*, *5*, 307 (1968).

83. W. M. Alvino, *J. Appl. Polymer Sci.*, *15*, 2521 (1971).

84. A. N. Goryachev and L. N. Pankratova, *Khim. Vys. Energ.*, *4*, 542 (1970).

85. Y. J. Kim and S. S. Lee, *New Phys. (S. Korea)*, *10*, 107 (1970).

86. Ya. M. Vus, N. D. Shcherba, and A. N. Tynnyi, *Fiz-Khim. Mekh. Mater. 6.*, 114 (1970).

87. P. Svoboda, *Anal. Fyz. Metody Vyzk. Plastu Pryskyric*, *1*, 230 (1971).

88. S. N. Zhurkov, V. I. Vettegren, I. I. Novak, and K. N. Kashintseva, *Dokl. Akad. Nauk SSSR*, *176*, 623 (1967).

89. S. N. Zhurkov, V. I. Vettegren, V. E. Korsukov, and I. I. Novak, *Fracture Proc. Int. Conf.*, *2nd*, *1969*, 545.

90. V. A. Kosobukin, *Mekh. Polim.*, *6*, 971 (1970).

91. V. I. Vettegren, V. E. Korsukov and I. I. Novak, *Plaste Kautschuk*, *19*, 86 (1972).

92. R. A. Dine-Hart, D. B. V. Parker, and W. W. Wright, *Brit. Polym. J.*, *3*, 222 (1971).

93. A. E. Portyanskii and Ts. M. Neiberg, *Vysokomolekul Soedin*, *10A*, 1394 (1968).

94. R. T. Sikorski, *Zesz. Nauk. Politech. Wroclaw. Chem.*, *18*, 65 (1967).

95. C. W. Strobel, U.S. Patent 3,646,142.

96. U. G. Gafurov and I. I. Novak, *Zh. Prikl. Spektrosk.*, *15*, 690 (1971).

97. B. Z. Volchek, A. I. Kol'tsov, T. N. Nekrasova, and A. V. Purkina, *Vysokomolekul Soedin*, *12B*, 754 (1970).

98. Y. Kyogoku, R. C. Lord, and A. Rich, *Biochim. Biophys. Acta*, *179*, 10 (1969).

99. J. P. Luongo, *Appl. Spectrosc.*, *25*, 76 (1971).

100. J. L. Koenig, *Appl. Spectrosc. Rev.*, *4*, 233 (1971).

101. S. W. Cornell and J. L. Koenig, *J. Polym. Sci.*, *7A-2*, 1965 (1969).

102.   J. L. Koenig, in *Applied Infrared Spectroscopy*, (D. N. Kendall, ed.), Reinhold, New York, 1966, pp. 245-284.

103.   J. L. Koenig and F. J. Boerio, *J. Chem. Phys.*, *50*, 2823 (1969).

104.   M. J. Hannon, F. J. Boerio, and J. L. Koenig, *J. Chem. Phys.*, *50*, 2829 (1969).

105.   T. Miyamoto and H. Inagaki, *J. Polym. Sci.*, *7A-2*, 963 (1969).

106.   T. Fujimoto, N. Kawabata, and J. Furukawa, *J. Polym. Sci.*, *6A-1*, 1209 (1968).

107.   R. Zbinden, *Infrared Spectroscopy of High Polymers*, Academic Press, New York, 1964, 264 pp.

108.   W. B. Rippon, J. L. Koenig, and A. G. Walton, *J. Agr. Food Chem.*, *19*, 692 (1971).

109.   V. I. Vettegren, I. V. Dreval, V. E. Korsukov, and I. I. Novak, *Vysokomolekul Soedin*, *12B*, 680 (1970).

110.   A. J. Melveger, *J. Polym. Sci.*, *10A-2*, 317 (1972).

Chapter 13

SURFACES

Clara D. Craver

Chemir Laboratories
Glendale, Missouri

I. INTRODUCTION

Analysis of the chemical structure of a surface is generally carried
out for one of three reasons: (1) to determine the composition of a
uniform bulk material, (2) to determine the composition of a surface

contaminant or layered coating, or (3) to investigate surface chemistry related to additive treatment, oxidation, radiation, or other exposure.

For some samples, all that is required is identification of the type of coating component that is the major constituent. This is usually easily accomplished by infrared (IR) or Raman spectroscopy because of their fingerprinting ability using any of several sampling techniques. For other samples, complete identification of all components is required. Quantitative procedures are necessary for this kind of application in order to establish that all key materials have been identified. The chemistry of interfaces in adhesive or catalyst research requires a still greater degree of experimental sophistication.

The techniques applicable to the analysis of coatings depend upon the nature of the coating material. Most commonly the coating is a formulated product made up of a few major components and several minor additives. Such a coating must be removed from the surface and fractionated into identifiable constituents for a reasonably complete analysis. A surface consisting of a single material may be analyzed in situ by internal reflection spectroscopy, by Raman spectroscopy, or by transmission spectroscopy by reflection if it is a transparent coating on a reflective substrate. Laminated coatings or coatings for which a composition gradient through the coating depth is to be investigated may require a combination of these procedures plus special techniques, such as KBr abrasion of successive layers, or microtoming.

It is significant that the term "compound" is avoided here. Many coating ingredients that are expected to be characterized as a single component are, themselves, complex mixtures chemically. Rosin, drying oils, waxes, asphalt, shellac, synthetic resins of varying molecular weight or differing in minor amounts of coreactants, or thermosetting compositions which cure as many different molecular species, all require identification as a broadly fixed but not completely reproducible mixture which must often be separated from other components equally broad in composition.

Thus, it is evident that the analysis of a coating may require a half hour if all that is needed is an identification of the general type of coating material such as "a wax," "an alkyd," or "principally polymer x." Otherwise, it may require many working days if each additive is to be quantitatively separated and identified or if the polymeric materials involved are modified by block or random poly- merization with other monomers and molecular conformation is required.

Special surface analysis problems may require data on chemical bonding or orientation of atomic groups within molecules at an inter- face. These needs are encountered in studying adhesives, the effect of primers or irradiation on surfaces, in catalyst research where the active surface may be a chemically treated fine powder, in treat- ment of paper or other fiber products to impart desired properties, and in biological research for studying the nature of membranes. A combination of physical methods is required to delineate the chemical structure of such interfaces.

Obviously, the spectroscopic and chemical techniques to be used depend upon which of these needs are to be filled and upon the nature of the surface itself. The objective of this chapter is to help the analytical spectroscopist understand clearly when a given technique is applicable and how best to obtain the results needed in research or technical service applications.

II. BACKGROUND AND LITERATURE

There have been thousands of scientific reports published in this field, including many excellent books. Accordingly, in any one chapter there is a major decision about how much to repeat old, valued examples, and how much to present details of newer or more difficult techniques. The choice has to include both.

In Chap. 1 (Part A of this volume), basic sample preparation tech- niques and spectral interpretation are discussed. The most useful gen- eral references are cited. In Chap. 12 on polymer identification, a schematic spectral interpretation section is fully described, and a comprehensive bibliography is presented on spectra of polymers.

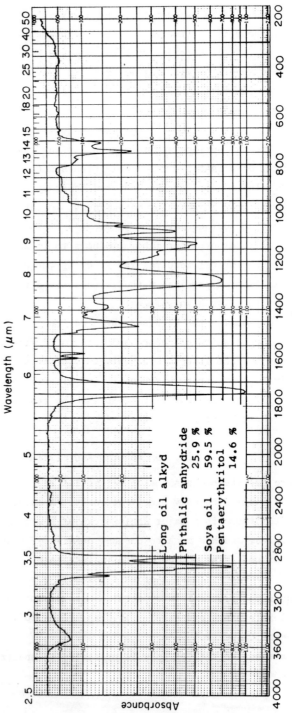

Long oil alkyd
Phthalic anhydride   25.9 %
Soya oil   59.5 %
Pentaerythritol   14.6 %

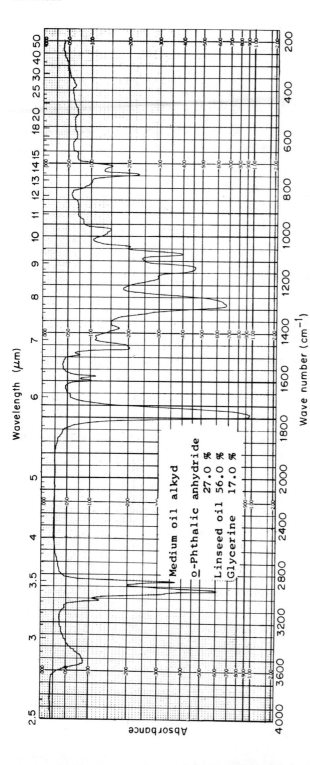

FIG. 1. IR spectra of phthalic alkyds. Differences between soya oil and linseed oil and between glycerine and pentaerythritol formulation components are inadequate to permit identification from a spectrum of the alkyd. Isolation of the oil and alcohol components from the ester is a required step for analysis. Reprinted from Ref. 1, p. 107, by courtesy of Federation of Societies for Paint Technology.

The most important references relating specifically to surface
coatings and examples of the kinds of published information available
are summarized below.

A. Spectroscopic Analysis of Coatings

Infrared Spectroscopy, Its Use in the Coatings Industry was prepared by specialists at DeSoto, Inc. in cooperation with the Chicago
Society for Paint Technology [1]. This book contains a survey of techniques and a reference library of 740 good-quality grating spectra,
including pigments, solvents, binders, and formulated paints. It is
invaluable for examples that demonstrate why good spectra and physical
and chemical fractionation are often necessary for a material to be
analyzed. For example, the two phthalic alkyds in Fig. 1 have only 26
to 27% of one starting material--phthalic anhydride--in common. As a
result, a large spectral difference might be expected. The phthalic
alkyd in the upper spectrum is made from soya oil and pentaerythritol,
and the one in the lower spectrum is made from linseed oil and glycerine. The spectra are so similar that only a carefully determined spectrum, in which band ratios and shapes can be relied upon and other
contaminants are known to be absent, reveals any difference at all.
Chemical isolation is required to distinguish between glycerine and
pentaerythritol and between soya and linseed oil. Study of the many
examples given in this book will help an analyst avoid the pitfalls
involved in giving too ready an interpretation of a spectrum. This
volume helps eliminate the time and costs of building up an extensive
reference library of spectra of known products to establish useful
ranges of spectral sensitivity to composition variations.

The spectra in this book cover the region 4000 to 200 cm^{-1}.
The low-frequency portion of the spectrum contains the most distinctive bands for many inorganic pigments. Figure 2 shows the ease
with which IR spectra distinguish between two forms of TiO_2, anatase
and rutile, and in Fig. 3 (page 940) the characteristic anion bands
for CrO_4^{2-}, SO_4^{2-}, and $Fe(CN)_6^{2-}$ are demonstrated. The Paint Federation
book contains 118 spectra of the most common pigments.

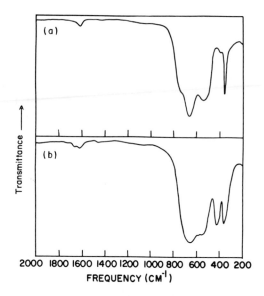

FIG. 2. IR spectra of paint pigments. Two forms of titanium dioxide, anatase (a) and rutile (b), are distinguishable from their IR spectra. Reprinted from Ref. 2, p. 464, by courtesy of Marcel Dekker, Inc.

Infrared Analysis of Polymers, Resins and Additives, An Atlas, [3] is in two volumes, with the first one so large that it is divided into *Part 1: Text* and *Part 2: Spectra.* This is the most comprehensive reference book on this subject area and it is well indexed. There are 1454 spectra of polymers and resins plus 320 spectra of monomers, solvents, and common resin and fiber components, such as the alcohols and acids of which many of the oils, plastics, and fibers are composed. These are prism spectra covering 2 to 15 μm or 2.5 to 18 μm. They are generally well run with multiple sample thicknesses and even with difference spectra included for cases where the interpretive data are thus improved. The identity of the materials sold commercially has been described more fully than is generally undertaken for mixed products. The most misleading bands, such as those from retained solvent, are identified.

As is the case with the Paint Federation book described above, quantitative data on spectra of common mixtures or chemical addition products are given. As the examples given in Fig. 4 show, it is

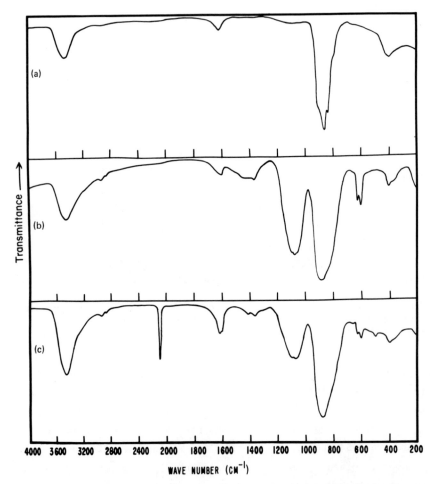

FIG. 3. IR spectra of paint pigments. Characteristic absorption
bands of anions permit identification in a mixed pigment: (a) chrome
yellow, medium ($PbCrO_4$); (b) chrome yellow, primrose ($PbCrO_4$, $PbSO_4$);
(c) chrome green, light [$PbCrO_4$, $PbSO_4$, $Fe(NH_4)Fe(CN)_6$]. Reprinted
from Ref. 2, p. 465, by courtesy of Marcel Dekker, Inc.

invaluable for the novice to see that between 12.5 and 13.5 μm there
are distinctive bands for identifying vinyl toluene- or styrene-modified epoxy resins. For the more experienced spectroscopist who would
already know the spectral characteristics for such major composition
differences, subtle help is available. It can be revealing to a

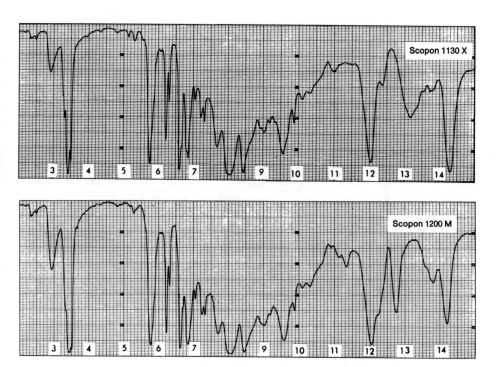

FIG. 4. IR spectra (transmittance versus μm) of epoxy ester resins modified by styrene (upper) and vinyl toluene (lower). Distinctive bands marked by arrows characterize the modifier. Reprinted from Ref. 3, by courtesy of Wiley-Interscience.

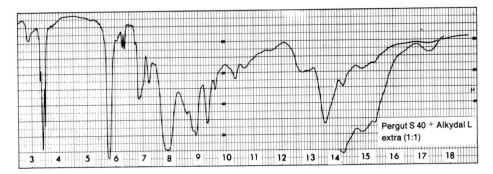

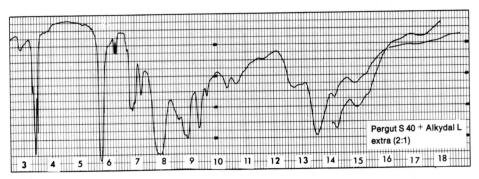

FIG. 5.   IR spectra (transmittance versus μm) of mixtures of alkyds and chlorinated rubber. A two-fold increase in chlorinated rubber content is difficult to detect by IR spectra. Reprinted from Ref. 3, by courtesy of Wiley-Interscience.

practiced spectroscopist that only small spectral differences exist between mixtures of chlorinated rubber and alkyd resin in which the proportions vary from 1:1 to 2:1 (Fig. 5).

B.   Combined Chemical and Spectroscopic Analysis of Coatings

A book which combines qualitative and quantitative chemical techniques with spectral data, *Identification and Analysis of Plastics* by J. Haslam and H. A. Willis [4], gives a useful series of spectra on vinyl chloride-vinyl acetate copolymers in varying proportions. These spectra show progressive spectral changes for polymers made from 98:2 to 84:16 vinylchloride/vinyl acetate. That there is

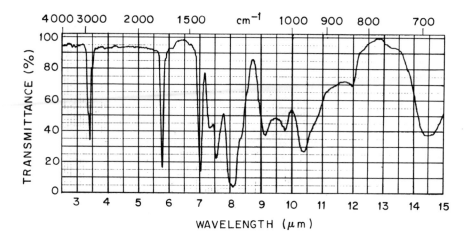

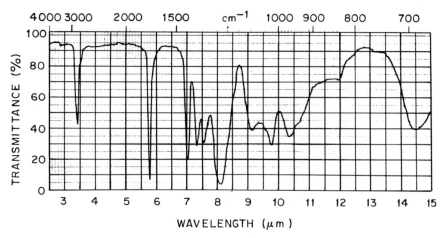

FIG. 6. IR spectra of vinyl chloride and vinyl acetate copolymers (upper, 95:5; lower, 89:11). Characteristic bands permit quantitative analysis of the copolymer ratio. Reprinted from Ref. 4, p. 381, by courtesy of D. Van Nostrand.

adequate spectral difference to identify only a minor change in these ratios is shown by the spectra in Fig. 6 of the polymers in the ratio of 95:5 and 89:11. The relative intensities of acetate bands (5.8

and 7.3 µm) to poly vinylchloride bands (7 and 14.2 µm) have been used as a basis for quantitative analysis. By spectral example (Fig. 7), this book warns of sensitivity difficulties in identifying small amounts of acrylic copolymers.

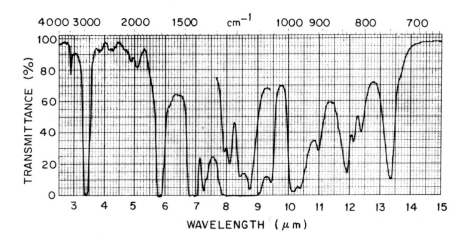

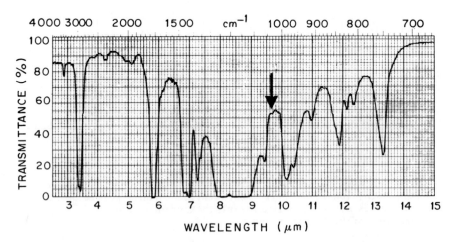

FIG. 7. IR spectra determined on thick films of polymethyl metha-crylate (upper) and methylmethacrylate ethylacrylate (lower) co-polymers. Modification of acrylic resins by a few percent of co-polymers may not be detectable in spectra determined at ordinary thicknesses. Reprinted from Ref. 4, pp. 392-393, by courtesy of D. Van Nostrand.

There are 25 spectra of hydrolysis products from resins plus other references grouped as vinyl resins (45 spectra), ester resins (31 spectra), nylon and related polymers (17 spectra), hydrocarbon and fluorocarbon polymers (32 spectra), rubber-like resins (24 spectra), and thermosetting resins (17 spectra). Other polymer categories add 22 more spectra and there are 44 spectra of plasticizers.

In addition to the reference spectra there are chapters on each of these classes of polymers which describe qualitative IR spectral characteristics and quantitative chemical procedures for their analysis.

Kappelmeier's *Chemical Analysis of Resin-Based Coatings* [5] is a classic volume on chemical separation and analysis of coatings materials. It includes procedures for spot tests, saponification, hydrolysis, or other chemical fractionation required to isolate co-reactants so that they can be identified positively. Equipment and methods described for such separations, such as vacuum depolymerization of acrylic resins or aqueous saponification of polyesters, yield much more reliable analyses than the much publicized short-cut procedures of pyrolysis. Kappelmeier's procedure for polyester analysis is outlined in Fig. 8 with recommended additions for control and follow-up by spectroscopy. An IR spectrum of the total sample before fractionation compared with spectra of the fractions prevents overlooking components which might not be readily recovered. Performing the separation on a weighed sample and obtaining weights of each fraction adds the control of a material balance to the procedure. Spectra provide proof of composition of each fraction.

The chapter on infrared spectroscopy in *Treatise on Coatings, Vol. 2* [2] by this author gives an introduction to the application of IR spectroscopy to practical coatings problems from the standpoint of describing to the coatings specialist the kinds of analytical data obtainable from IR spectra. It provides elementary examples which demonstrate the spectral additivity of functional groups within polymers for general identification (Fig. 9) and describes applications which involve careful interpretation of small spectral differences.

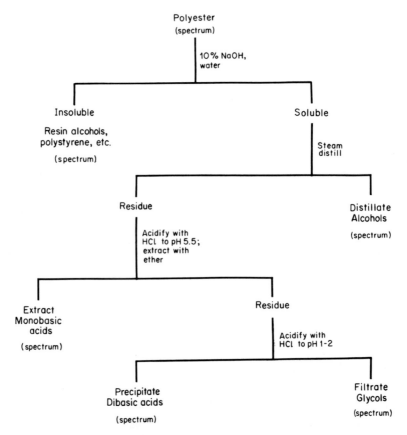

FIG. 8. Flow diagram for combined chemical and spectroscopic analysis of a polyester. Adapted from Ref. 5, p. 572, by courtesy of Wiley-Interscience.

Standardized analyses are described in detail in the ASTM series, *Annual Book of ASTM Standards* [6], Part 20: Paper, Packaging; Part 21: Cellulose, Flexible Barrier Materials, Leather; Part 28: Pigments, Resins and Polymers; Part 29: Naval Stores, Aromatic Hydrocarbons; and Part 27: Paint--Tests for Formulated Products and Coatings. ASTM's *Manual on Recommended Practices in Spectrophotometry* [7] covers instrumentation and general techniques. In 1974 a manuscript was prepared to expand this book to include detailed recommended practices for internal reflection spectroscopy [8]. Other

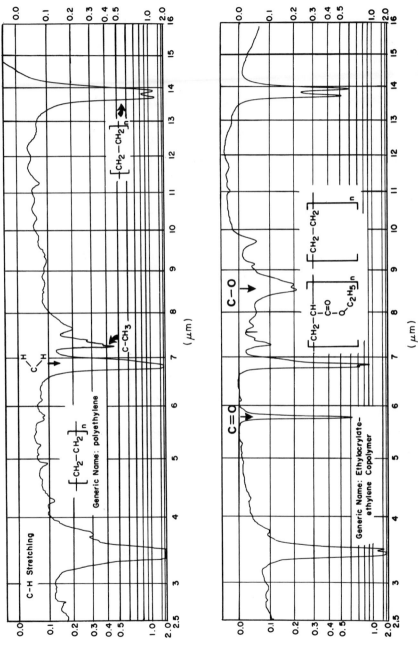

FIG. 9. Comparison of the IR spectra of polyethylene and ethylacrylate-ethylene copolymer demonstrates the spectral additivity of functional groups characteristic of each monomer. Reproduced by courtesy of R. A. Nyquist.

volumes available from ASTM pertain to composition and analysis of
aerospace materials and waxes. ASTM methods provide valuable committee recommendations that have industry-wide acceptance for procedures which combine general analytical tests, fractionation, and
spectroscopy.

The Gardner/Sward *Paint Testing Manual* [9] suggests composition
possibilities to the analyst. It covers typical film formers, solvents, natural and synthetic resins, waxes, and complex formulations.
The chapter by Swann [10] provides selected methods which combine
analytical chemistry, gas chromatography (GC), and spectroscopy for
the analysis of the most important commercial coating resins by
methods that he considers most satisfactory. These are detailed
methods and include the analytical wavelength, the procedure for
calibration, and examples of calculations. The procedures are inter-
spersed with references to appropriate ASTM methods.

C. Components of Coatings

Books on the technology of paints and plastics are of considerable utility to the analyst working in the field of applied coatings
because he needs to know physical properties of coatings components and types of compatible formulations. Such books include
the annual *Modern Plastics Encyclopedia* [11], *Treatise on Coatings:
Film-Forming Compositions* by Myers and Long [12], a series of 21
booklets published as units in the *Federation Series on Coatings
Technology* [13], and two volumes published annually as *Coatings and
Plastics Preprints* [14] (papers presented at National ACS Meetings
by the Division of Organic Coatings and Plastics Chemistry of the
American Chemical Society).

The kinds of assistance that can be expected from these volumes
are demonstrated in Tables 1 and 2. In Table 1 is the formulation
for an exterior acrylic paint. The problems its composition present
analytically are evident, and the list of typical components aids the
analyst in planning an analytical scheme.

TABLE 1

Formulation of Gloss-baking Enamel with
Thermosetting Acrylic Emulsion[a]

Material	Pounds/100 gal
Rutile titanium dioxide	200.0
Water	80.0
Amberlac 165 (22% solids)(Rohm & Haas)	7.0
Tributyl phosphate	1.0
Dimethyl amino ethanol	2.0
Rhoplex AC-201 (46%)(Rohm & Haas)	668.0
Dimethyl amino ethanol (50% in water)	15.0
Rhoplex B-15 (46%)(Rohm & Haas)	52.0
Butyl Cellosolve (60% in water)(Union Carbide)	58.0
Tributyl phosphate	1.0
Total	1084.0

[a]Abridged from Ref. 15, p. 30, by courtesy of Marcel Dekker, Inc.

Another way that these nonspectroscopic books are helpful is in suggesting compatible components in a particular kind of product. For example, a paint which gives exceptionally good adhesion may be found by preliminary analysis to be based on chlorinated rubber. A list of other probable components, including data which relate its composition to substrate, is available from a formulation table (see Table 2).

A method which will be discussed several times in this chapter is the use of solvent extraction to clean up a polymer so that it can be precisely identified. As valuable as this technique is, it can lead down a long path of isolation of unidentifiable minor components. This is especially so if one of the constituents of the coating is a natural gum or resin. These resins have exotic names like accroides, damar, batu, elemi, and sandarac and contain components such as high-molecular weight hydrocarbons, alcohols, acids,

TABLE 2

Resins That Give Good Adhesion with Chlorinated Rubber[a]

	Aluminum	Steel	Wood	Concrete	Cellophane
Resin					
Long-drying oil alkyd	x	x	x		
Chlorinated diphenyls	x	x		x	
Coumarone resins	x			x	x
Acrylic resins		x	x	x	x
Chlorinated paraffins		x		x	
Plasticizers					
Nondrying alkyds	x	x	x		
Chlorinated diphenyls		x		x	
Chlorinated paraffins		x		x	
Dioctyl phthalate	x				x
Linseed oil		x	x		
Tricresyl phosphate	x		x		
Desavin		x	x	x	
Polyvinyl methyl ether		x		x	x

[a]Abridged from Ref. 15, p. 30, by courtesy of Marcel Dekker, Inc.

and esters. Compounds have been identified from C_{10} to at least C_{65} with the highest concentration for most resins being in the C_{25} to C_{45} range. The number of individual compounds in one resin may be very high. Obviously with such a variation in molecular weight and chemical functionality these compounds from a single coating constituent can divide among many fractions as classical separation procedures are carried out.

Resins such as these impart highly specific properties to coatings. To consider one, elemi, defined as "any of various resins from certain trees esp. *Camarium commune*," is described by Mantell [15] as a plasticizer that confers increased adhesion to lacquers applied to metal surfaces. He lists products in which it is used as adhesives and cements, wax compositions, printing inks, surface coatings applied to textiles and paper, linoleum and oilcloth, as a base in perfumes, and as an ingredient in fireproofing and water-proofing compositions as well as in engraving and lithography. It is further described as compatible with vegetable, fish, and animal oils of both drying and nondrying types, waxes of mineral and vege-table variety, as well as stearic acid, cellulose derivatives, phenol-aldehyde resins, maleic rosin resins, coumarones, petroleum, asphalt, coal tar, and others. It serves as a mutual solvent in resin-wax combinations.

This high compatibility with different components in a coating may make elemi difficult to separate for identification. If sepa-rated, its spectrum (Fig. 10) is not particularly informative except for fingerprint comparison. Mantell lists 27 known constituents of elemi, with some of these identified only as a class of compound.

The analyst who does not familiarize himself with the expected components in coatings can inadvertently let an analytical fraction-ation develop into a research project. These books are one of the best substitutes for that greatest of all possible helps in performing an analysis: a specialist who knows what the likely components of any given system are.

III. ANALYSIS OF SURFACE COATINGS

This section is a critical and selective evaluation of the many al-ternative experimental approaches to obtaining chemical structure characterization of the composition of surfaces and surface coatings. It is well known that molecular spectra offer the most distinctive identification data possible for compounds and complex mixtures. The greatest difficulties in practical analyses arise in obtaining pure enough samples for making possible unambiguous identification.

As already described, most organic surface coatings are complex formulations consisting of binders of a resinous, plastic, or rubbery material, plasticizers, stabilizing ingredients, and fillers or pig-ments. An analytical request might be filled by a report on the type of major components present, or it might require qualitative identi-fication and quantitative analysis of all components.

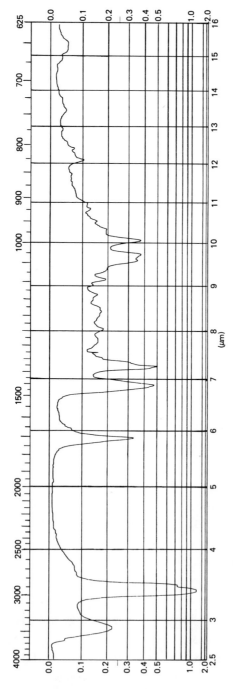

FIG. 10. The IR spectrum of elemi [trade name: Gum Elemi (natural resin); supplier: S. Winberbourne] has bands at OH, C=O, and CH_2, CH_3 frequencies common to most resins and many polymers. This natural product composed of many compounds is not readily identified in a formulated coating. From Coblentz Society Spectra Vol. 10, by permission of the Editor.

A perspective in practical analysis, if it is also to be a reliable analysis, requires that some of the ways that wrong answers are obtained by given attention. There is no journal of negative results, and instruments and sampling devices are sold on the basis of publicity about what they will do and not what they will not do. This point can lead to an overly optimistic view about the amount of experimental caution that is required for reliable analytical data. Despite the enormous power of IR and Raman spectra for finger-printing molecules, sound analytical procedures must be followed in sampling and sample purification.

A. Identification
    of Major Components

   1. Sample Preparation
   a. Bulk Samples. The simplest coating material to analyze is the sample before it is applied. If it is a solid polymer for hot melt or laminating operations, it can be prepared for spectral deter-mination in any of several ways: a small amount can be melted in a thin film on a transparent plate for transmission spectra or on an internal reflection element, or the polymer can be dissolved and cast as a thin layer directly on the sample plate or onto a glass or mercury surface and stripped off as a free film.

   Spectra of thin, unsupported films can be run directly by trans-mission in a thickness range of about 0.01 to 0.2 mm. Most materials give good qualitative spectra at less than 0.1 mm, but thicker films may be useful for detection of weak bands in a nonpolar material such as a nonaromatic hydrocarbon polymer.

   A formulated solvent varnish or other clear coating may be run as a film cast from the solvent on a sample plate. Solvent-based paint may likewise be cast as a film, but interference from pigments or fillers frequently is too great for adequate qualitative inter-pretation of the spectrum. For pigmented coatings it is desirable to separate the organic binders from the inorganic matter. Dilution and centrifuging is usually a successful way of effecting this sepa-ration.

Latex paints may be prepared as free films to eliminate water, which attacks common sample window materials. They will form a film on a loop of wire dipped into the paint and allowed to air-dry. The success of this technique depends upon the properties of the paint. For some paints an exceptionally fine wire and small loop of about 1/2 in. diameter may be needed in order for the film to air-dry without breaking. Casting a film on a glass plate and then peeling off the free film after it has dried may be easier than forming a film in a wire loop for some samples. Water solutions may be put directly on some kinds of sample windows and allowed to dry.

McGinness [16] has provided a good description of general techniques for analyzing whole paint. He includes an excellent discussion of how to obtain representative samples.

b. Applied Coatings. An applied coating may be dissolved off the surface, scraped from the surface, or, if the coated specimen is small, reasonably flat, and on a slightly yielding substrate, it can be put directly into an internal reflection attachment.

Some simple procedures for preparing a whole coating for preliminary qualitative spectra are pictured in Fig. 11. (1) A soft transfer coating is removed from paper with a razor blade to avoid contamination from the substrate or the coating on the other side, (2) a soluble paper coating is washed on one side with a minimum of solvent to minimize contamination from the other side, (3) a surface is abraded with KBr, (4) a small amount of material at a junction is removed by softening with a solvent and pressing a sample plate against it.

Wire coatings are sometimes soluble but more often they are highly cured and difficult to remove. Dimethylformamide is an effective solvent for many wire coatings. Cured wire coatings may be released from the wire by dipping them into concentrated sulfuric acid for approximately 1 to 3 min. The coating disintegrates if it is left in the acid too long, but if it is removed from the acid as soon as loosening is observed, and rinsed in water, pieces of the coating can be unrolled and placed directly on a sample plate. Very

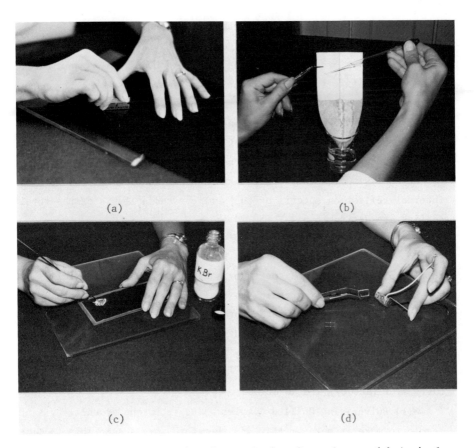

FIG. 11. Sample preparation for analysis of coatings. (a) A single-edge razor blade scraped across a transfer paper backed by a glass plate. (b) A coating on one side of paper is washed off by trailing solvent from a microdropper across the top of the sheet held vertically and catching the solution in a beaker. (c) The top surface of a weathered asphalt film on a test panel is removed by gently abrading the top surface with KBr. Light scraping with a razor blade also effects removal of the top layer. (d) A polymeric speck in an electrical system is sampled by wetting it with solvent and lifting it with a small sample plate that fits into a microholder.

fine wire might be most easily prepared for a spectrum by lining up the wires on an internal reflection plate. Pyrolysis (discussed in the next section) is also used for wire coatings to determine components of the major resin.

These examples demonstrate that sample handling for qualitative
IR spectra is highly varied and is dictated by the nature of the
sample. Spectra of samples from any of these preparations could be
obtained by either internal reflection or by transmission. For paper
samples, internal reflection spectra can be used to analyze either
side of the sample without obtaining an interference from materials
on the other side. An internal reflection spectrum of the surfaces
does not necessarily give a spectrum of the bulk of the coating.
For an analysis of a completely unknown paper coating system, both
internal reflection spectra of the surfaces and analysis of the total
coating are desirable.

      c. Intractible Samples. The kind of sample that is most diffi-
cult to prepare for an IR spectrum is hard, insoluble, infusible,
not friable enough for grinding to small particles, and in irregular
or curved shapes that are not adaptable to close contact with internal
reflection elements. This describes highly cross-linked polymers
ranging from elastomers to casting resins. It includes also some
baked industrial finishes. Three common avenues of approach, none
fully satisfactory, are described below.

      (1) Pulverizing Chilled Samples. The technique of chilling a
sample to make it more brittle and therefore more susceptible to fine
grinding is well known. A device called the Freezer/Mill, available
from SPEX Industries, Inc., Metuchen, N.J., has made this procedure
straightforward. The sample is placed in vials with magnetic end
caps but with nonmagnetic center sections. Coils are alternately
energized and a rod-pestle is driven back and forth between the caps
at up to 30 times/sec to grind the sample. This grinding unit is
surrounded by a bath for coolant adapted to handle liquid nitrogen.
This cooling embrittles the sample and provides fine grinding. It
also prevents development of hot spots in sample handling and possible degradation of heat-sensitive compounds. A spectrum of the
finely ground sample can be run as a mull, a pressed halide disk, or
a powder film by either transmission or internal reflection.

(2) Raman Spectra. Raman spectroscopy offers a method for handling a tough, intractible, physically irregular sample except that its application is limited by sample fluorescence. Raman spectra have been reported for a variety of physical shapes [17], as shown in Fig. 12. Raman spectra from these samples are useful for identification of the major polymer material present, as may be seen by comparison of the spectra of the polycarbonate resin and a poly methylmethacrylate resin in Fig. 13.

Obtaining Raman spectra of commercial products is still an uncertain process because of fluorescence, which can dominate the spectrum. Sampling systems are discussed on pages 971 and 972.

(3) Pyrolysis. Most laboratories do not yet have Raman capability, and a more common approach to determining spectra of intractible polymeric material is by pyrolytic degradation to form identifiable gas and liquid fractions and/or a soft residue. Difficulties in obtaining reproducible pyrolyzates have been pointed out repeatedly

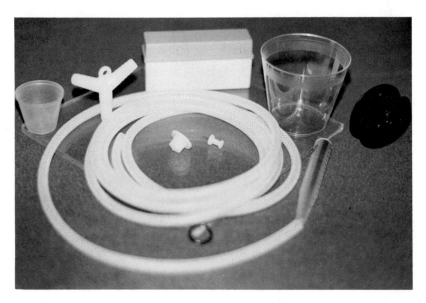

FIG. 12. Plastic materials of such diverse shapes as test tubes and buttons can be mounted directly for determining Raman spectra of the surfaces. Reprinted from Ref. 17, p. 33, by courtesy of Plenum Press.

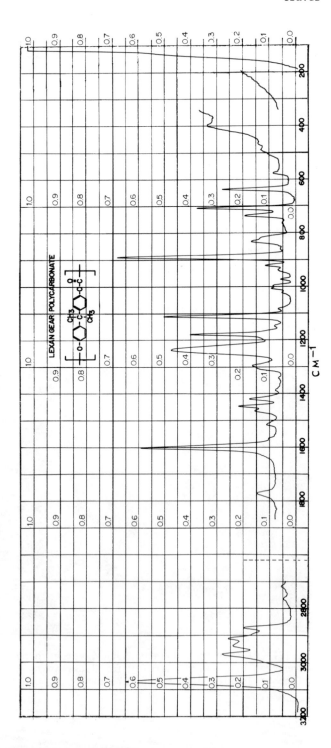

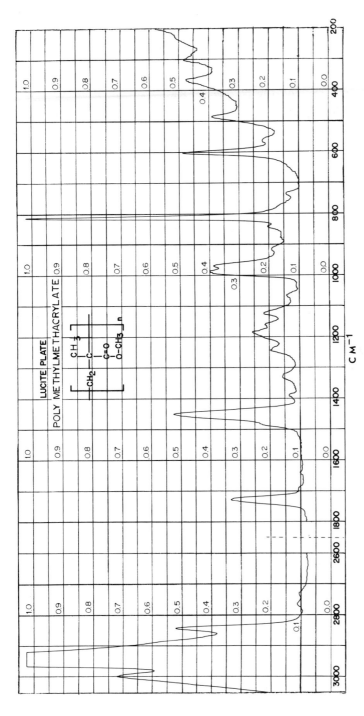

FIG. 13. Raman spectra of polycarbonate gear (upper) and poly methylmethacrylate plate (lower) in Fig. 12. Reprinted from Ref. 17, p. 3435, by courtesy of Plenum Press.

by experimenters, especially by those obtaining GC spectra on the
pyrolyzate. They desire a pyrolysis product so uniform that a
chromatogram of the pyrolyzate can serve as a fingerprint [18,19].
Pyrolysis of polymers may yield fragments similar to either a low-
molecular weight form of the polymer product [20] or other identifiable decomposition products. The analytical problem is not whether
the fragment is a known, identifiable substance, but whether it
represents the total sample, or whether in fact it is truly interpretable at all as a part of the sample in its original state.

A recent sample would have defied most standard pyrolytic procedures unless all fragments over a wide volatility range were
identified. Even then the original nature of the sample would probably have been misinterpreted. Since this sample was erroneously
analyzed by a well-known commercial laboratory using a standardized
pyrolysis and GC procedure, it will be described here as a warning
of a major pitfall of the pyrolysis technique.

Case History: An opaque, white plastic molded cylinder for use under
stringent performance conditions was being analyzed so that routine
materials control testing could be set up. The cylindrical shape
and tough appearance of the plastic suggested pyrolysis as the method
of choice to the laboratory which first performed the analysis. An
abundance of melamine in the pyrolyzate was readily identifiable, and
the sample was reported as a melamine resin, possibly with some silica filler, because Si was identified in the ash.

This analytical report did not agree with what was known about
the manufacturing properties and performance characteristics of the
product, and a second analytical laboratory was consulted. Straightforward quantitative analysis consisting of separating fractions,
identifying them, and weighing them gave a far different answer.

Solvent washing of the surface isolated a trace coating of
silicone oil which was the source of the reported Si. More than 60%
of the sample was found to be polymethyl methacrylate resin, with a
few percent of a common phthalyl glycolate, and an identifiable ester
wax and oil. The white "filler" in this specialized product was

about 36% of the total sample and was isolated as residue from the extraction. It was pure melamine--the compound, not the resin. This melamine had vaporized during heating and had been identified as being a decomposition product of melamine resin during the earlier "pyrolysis" analysis. The polymethyl methacrylate had not even been found. A more careful pyrolysis analysis would surely have identified the PMMA, but chances are that the compound melamine would have been reported as a resin in most laboratories relying on pyrolysis data alone.

As a matter of interest a Raman spectrum of this sample was obtained. It fluoresced so badly, even after a 4-hr exposure to the laser source, that only weak lines could be observed at three of the strongest polymethyl methacrylate frequencies.

Procedures which miss 60% of the sample, as in the above example, should be avoided. Most experienced spectroscopists are wary of pyrolysis and either use it only as a spot check on routine samples, or are thorough in its use and combine it with GC followed by IR or mass spectroscopy. A quick pyrolysis "test" in which one or two pyrolysis fractions are found to correspond to a known should not be upgraded to the status of an "analysis" in an analytical report.

2. *Spectrum Determination*

a. Infrared Transmission. The nature of the sample determines the physical preparation necessary for a transmission spectrum. If there is an abundance of material and if it is a readily soluble film former, casting it on a glass plate and peeling if off provides sample areas of different thicknesses for determination of a spectrum on both a thick and thin specimen.

The spectrum of the thin specimen is useful for comparison of band intensity ratios of strong functional group bands such as C-H, C=O, Si-O, etc., of the major components. The spectrum of the thicker specimen may be necessary to detect copolymers; see, for example, the weak copolymer bands in Fig. 7.

The major source of error in the spectrum of a solvent-cast film is retained solvent. Therefore, whenever possible, a solvent should be selected that will have distinctive bands in the spectrum of the sample. For example, residual acetone in a polymethyl methacrylate spectrum may be most visible as a shoulder on the $C-CH_3$ band and leave the spectral interpretation in doubt. The addition of benzene to the sample film can release the entrained acetone. Residual benzene can be detected by its band at 680 cm^{-1}, which is such a distinctive frequency that it seldom interferes with polymer identification.

The one or two strongest peaks of any solvent used for a cast film should not be overlooked in a spectrum. It is preferable to inspect the curve while the spectrum is being determined, so that further drying at that time will permit subsequent interpretation of the spectrum without solvent interference.

Minor problems in interpretation of the spectrum of a free film arise from interference fringes and from quantitatively uneven distribution of components within the film. The interference fringe problem is not likely to be important except for commercially produced films for which the surfaces are parallel. Even for commercial products interference fringes do not prevent qualitative analysis. Interference fringes are a source of error in quantitative analysis not only for films but for liquids in fixed thickness cells.

The magnitude of the quantitative error from uneven distribution of components in a cast film is a function of both the structure of the components and their molecular weight, as well as their flow characteristics in the particular solvent being used. The rate and uniformity of drying and other factors which control film creeping determine the homogeneity of a cast film. Extensive tests for repeatability of film deposition and selection of the optimum conditions are necessary for quantitative analysis of a mixture.

The ease of film preparation for qualitative work makes it a widely used method either as the free film technique or as a film deposited on a sample window. Surprisingly small samples can be

handled by deposition from solution, but a bit of skill is involved.
The concentration of the sample in the solvent, the surface tension
characteristics of the mixture and of the drying polymer film, and
the rate of addition of the solution to the plate relative to the
solvent evaporation rate determine whether a reasonably uniform film
will coat the plate or whether the residue from evaporation will be
carried to the edge of the plate. (See Fig. 14.)

Pellet (halide disk) and mull preparations are widely described
and will not be repeated here. The important hints for obtaining
good spectra by the oil mull technique are to grind the sample fine
and to add a minimum of oil [21]. Freeze-dry methods are highly
recommended for preparation of pressed halide disks. It is this
author's experience that interpretation of spectra of pellets adds
enough uncertainties to qualitative interpretation of an unknown
sample for it to be the method of choice only when ultramicrosamples
are involved, or for routine quantitative analysis of known products.

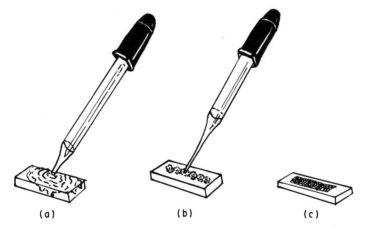

(a)   (b)   (c)

FIG. 14. Deposition of polymer films from solution may be adapted to
very small samples: (a) An oversized sample plate, too much solvent,
and too rapid solvent flow force the polymer to an area outside of
the light beam and even over the edges of the plate. (b) Use of a
sample plate just slightly larger than the beam area, a warm sample
plate, a warm solution, and a microtipped dropper with the solution
allowed to dry as it is deposited permits alignment of minimal sample
in the light path with little sample loss. (c) The sample can be confined in a small area scratched on a salt plate.

Educational films on sample preparation techniques have been prepared under the auspices of the Education Committee of the Coblentz Society, Inc. [22]. An 8-mm movie film on mulling technique is available from the Society on loan free of charge, and a slide series with sound is available for purchase from Communication Skills, Inc., Fairfield, Conn.

   b. Internal Reflection Spectroscopy. The important variables in internal reflection spectroscopy have been discussed in Chap. 1. A discussion of theory and a summary of the field to 1967 has been written by Harrick [23]. In internal reflection spectroscopy an optical attachment to the spectrometer is used to direct the sample beam through a transparent internal reflection element (sample plate) against which the sample is placed. An angle of incidence is selected which permits the radiation to traverse the element and be reflected from the element-sample interface a selected number of times, and then ultimately be redirected into the spectrometer. Some of the energy of the radiation penetrates the sample at the point of reflection, with the absorption of energy occurring in the same way as in transmission spectroscopy.

   The geometry of the internal reflection optics and the relative refractive indices of the sample and the internal reflection element determine the number of interface reflections and the depth of penetration of the sample beam and, therefore, the absorption intensity of a given sample. Depth of penetration of the wave is also a function of the wavelength of the radiation. A greater depth of sample is penetrated at longer wavelengths than at shorter wavelengths. Therefore, the absorption characteristics of a nonuniform stratum of sample is measured at successive intervals in the spectrum.

   Practical application of internal reflection spectroscopy involves the control of these variables. The spectra obtained may be quite similar in appearance to transmission spectra or differ considerably in band shape, intensity and, to a lesser extent, position.

For versatile applications to surface analysis, internal reflection attachments which permit a selection of angle of incidence and a choice of internal reflection elements adequate to provide up to 40 reflections on materials ranging from low to high refractive indices are needed. An equivalent reference beam unit is, of course, often desirable.

Instructions for optical alignment and tests of energy throughput of internal reflection attachments are provided by the manufacturers and should be followed carefully to assure satisfactory performance of the system.

The choice of material for an internal reflection element and the quality of maintenance of the element determines to a large degree the quality, and therefore the usefulness, of an internal reflection spectrum. Properties of common materials for internal reflection elements are given in Table 3. KRS-5 is suggested as the material with the most favorable combination of characteristics for general organic applications. For samples for which a low refractive index element is acceptable, the low cost of silver chloride provides an advantage.

The selection of an internal reflection element for a particular sample is ultimately determined by trial and error. A region of the spectrum is selected in which no bands are expected to appear, e.g., 5.0 to 5.5 μm, and the system of internal reflection attachment, sample plate, and sample is adjusted to give high energy throughput. Then a manual or high-speed spectral scan is made to ascertain absorption band intensities. If the band intensity (spectral contrast) is not adequate for spectral identification, adjustment of surface area, contact pressure, angle of incidence, or a change to a different internal reflection element may improve the spectrum. Minor adjustments of band intensity may be made by stopping the instrument on an absorption band and watching the pen deflection as adjustments are made. It is necessary, of course, to recheck a nearby absorption minimum to establish that increased spectral contrast, not a loss of total energy throughput, is achieved by these adjustments.

TABLE 3

Properties of Typical Optical Materials for Internal Reflection Elements[a]

Material	Useful Range (μm) 15-mm path	Mean refractive index n_1	Critical angle O_c(°)	Comments
Silver chloride	0.4-20	2.0	30.0	Very soft, moldable; easily scratched; light-sensitive; least expensive
Silver bromide	0.45-30	2.2	27.0	Slightly harder than AgCl, otherwise similar
KRS-6 (thallous bromide chloride)	0.4-32	2.2	27.0	Similar to KRS-5; expensive
KRS-5 (thallous bromide iodide)	0.6-40	2.4	24.6	Toxic[b]; moderately soft; convenient refractive index; most favorable combination of characteristics
Silicon	1.1-6.5	3.5	15.6	Hard; high surface polish; high resistivity; useful at high temperatures to 300°C
Germanium	2.0-12	4.0	14.5	Limited spectral range; sensitive to temperature; opaque at 125°C

[a]Reprinted from Ref. 8 by courtesy of the American Society for Testing Materials.

[b]Toxicity is not too great for ordinary handling of polished element; fine particles from grinding and polishing should not be allowed to come in contact with the skin.

The quality of an internal reflection element depends on the quality of the optical material and the precision of its manufacture. Some materials, especially KRS-5, may exhibit a hazy condition internally which will scatter short wavelengths. The manufacturing tolerance of all angles and surfaces of an internal reflection element through which light is transmitted, or from which reflection occurs, is strict.

Internal reflection elements should be tested in a well-aligned attachment for high total transmittance, uniformity of transmittance level throughout the spectral range, and absence of extraneous absorption bands. Evaluation of both new and used internal reflection elements is expedited by retaining an unused high-quality element as a comparison standard.

Internal reflection elements are easily damaged. Bent, dented, hazy, or scratched surfaces do not give satisfactory reflection at the surfaces or transmission through the element. For cleaning the elements it is recommended that solvent washing and a gentle dabbing motion with clean, soft rayon balls be used. Coets® have been found to have negligible extractable material [8], and the soft inner surface is excellent for cleaning internal reflection elements.

Self-adhering samples are as easily applied to the internal reflection element as they are applied to transparent plates for transmission spectra. Pastes, resins, and paints can be smoothed on directly and soluble solids can be deposited from solution. Care must be taken to avoid scratching the crystals. The solvent must be removed by evaporation. This step of solvent removal may be easier for samples on internal reflection elements than on transmission plates, because a thinner film can be used and the problem of solvent entrapped by a dried surface layer of resin is minimized. (To obtain transmission spectra of very thin films which might be needed for materials for which solvent retention is a major problem, multiple surfaces of cell windows can be used; thin films can be deposited on several plates and the windows stacked to form a single transmission unit.)

Nonadhering samples such as films, fibers, powders, or coated surfaces must be pressed against the internal reflection element and held in optical contact with the crystal throughout the spectral determination. Pressure must be uniformly distributed across the crystal or breakage or bending may result. Too little pressure can result in incomplete contact of the sample with the optical element. Too much pressure can cause fine scratches or depressions in the crystal if particulate matter is present in the sample or if the sample is hard or unyielding or not optically flat.

A soft backing of rubber or Teflon® between the sample plate and the metal backing plate helps distribute the pressure uniformly across the crystal. Pressure can be controlled and reproduced from one sample to another by use of a torque wrench to tighten the screws on the back pressure plate. Care must be exercised to have the sample cover the entire area of the element exposed to the backing cushion, or absorption bands from the cushion will appear superimposed on the spectrum of the sample.

Fibers or fine wires should be carefully aligned on an internal reflection element without overlapping or crossing. This care in placing the sample on the plate is important if the pressure of the backing plate is to be applied uniformly enough to give a reasonable amount of coverage of the sampling area and if indentations are to be avoided on the soft elements. One technique for this alignment is to place the filaments on an adhesive or masking tape before placing the assembly against the element. A spectral data blank should be run on the tape so that it will be recognized in the spectrum if the adhesive migrates to the element sample interface and gives rise to bands that are not representative of the sample.

In internal reflection spectroscopy only the surface in contact with the optical element is measured. Thus an internal reflection spectrum offers an opportunity to record data on a top surface onto which some component of a coating has migrated or which has been subjected to degradation from exposure to oxidation or chemical treatment. This same phenomenon can become a disadvantage if the desired information is the identity of a polymer and the portion of the sample on the surface is the plasticizer of the system. Polyvinyl chloride resins have been identified as phthalic alkyd coatings because they were plasticized with a phthalate ester, and the fluid ester exuded under pressure and wet the internal reflection element. In this situation the major portion of the sample, the polyvinyl chloride resin, may lie completely beyond the depth effectively penetrated by the light beam and may not be detectable in the spectrum at all.

Internal reflection spectra offer an advantage over transmission spectra for microgram sample sizes. Not only can multiple reflections be used, but also double-pass elements are available which increase the effective pathlength still further.

To make optimum use of the surface area of an internal reflection element for microsamples and to obtain repeatable band intensities for quantitative applications, it is necessary to test the internal reflection elements in any given optical arrangement to determine the most sensitive areas--that is, the area where the light beam is most concentrated as it reflects from the interface between element and sample [24].

This test can be made several ways. One method is to move an opaque object, such as an index card, along the beveled edges of the internal reflection element until the transmitted energy is decreased. This procedure will define one dimension of coverage of the internal reflection element by the light path.

A more quantitative calibration of a reflection element can be made by testing the intensity of an absorption band as incremental areas of the surface are covered by a sample which readily coats the surface. Alternating and erratically variable regions of sensitivity may be detected on an internal reflection sampling face, as shown in Fig. 15. Typical commercial internal reflection elements are highly variable and display regions of tremendous sensitivity differences. The way an element is positioned in the light path can also affect the sensitivity of an area of its surface. An example is given in Fig. 16 where the same sample on the same plate in the same optical

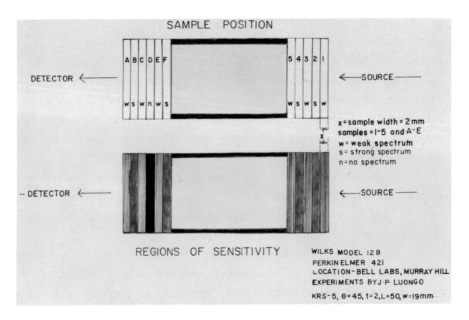

FIG. 15. Strips across the sampling face of an internal reflection element show high variability in sensitivity of surface areas. No spectrum at all was observed at the strip marked D. Reprinted from Ref. 25, by courtesy of C. Paralusz and J. Luongo.

reflection attachment was simply changed in the holder so that the exit bevel was interchanged with the entrance bevel. It is obvious that this variation in sensitivity from one portion of the internal reflection element surface to another can mean the difference between success and failure in obtaining a spectrum of a microsample which cannot be spread over the entire surface.

For quantitative applications it is important to rely on band ratios or to obtain the calibration data with the same plate that is used for the sample, and to test the geometry of the plate for extraordinary sensitivity excursions as it is repeatedly replaced in the sample holder. For a soft, flowable sample that readily coats the entire surface of an internal reflection element, sample sensitivity irregularities are averaged over the surface. Even for samples smaller than the internal reflection element, it is possible to obtain spectra with reproducible absorption intensities and, therefore,

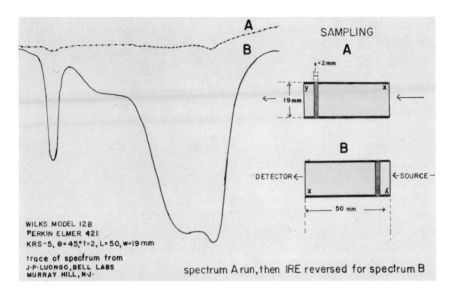

FIG. 16. An internal reflection element at reversed positions in a reflection attachment varies from producing a strong spectrum to a barely detectable spectrum. Reprinted from Ref. 25, by courtesy of C. Paralusz and J. Luongo.

to perform quantitative analyses. Successive samples must be placed in identical positions on the internal reflection element. In addition, the optical condition of the element and optical alignment of the internal reflection attachment must not change between calibration and analysis. Band ratio methods are usually more reliable than intensity measurements, which depend on complete experimental repeatability.

Very thin films on an internal reflection element yield spectra more similar to transmission spectra than do bulk samples. Spectra of thin films thus have the advantage that the large reference catalogs already available for transmission can be used for comparison with the internal reflection curves for identification of major components.

c. *Raman Spectroscopy.* Typical sample mountings used to obtain Raman spectra are described in other chapters in this volume and are discussed by McGraw [26]. A schematic depiction of front surface laser reflection and laser transillumination is given in Fig. 17.

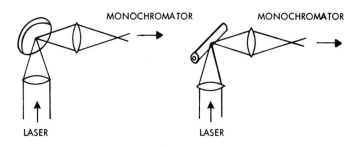

FIG. 17. Diagrammatic depiction of front surface sample mountings
for determination of Raman spectra. Reprinted from Ref. 26, p. 38,
courtesy of Plenum Press.

The greatest experimental problem in determining Raman spectra
of commercial materials is sample fluorescence, which can conceal
the relatively weak Raman spectrum. Although some commercial samples
have little enough fluorescence that a laser Raman spectrum can be
readily obtained, others may fluoresce badly.

Cleaning up the sample whenever possible is the optimum way to
minimize or eliminate fluorescence. A sample that fluoresces with
one laser source may not fluoresce with a laser of different frequency.
Several investigators have reported that exposure of some samples to
a laser beam for times up to several hours may exhaust the fluorescence and permit a good spectrum to be obtained.

3. Spectral Interpretation

a. *Infrared.* Preliminary IR interpretation of a spectrum of
a coating mixture is a valuable guide to subsequent chemical and
physical procedures, and even suggests the kinds of additional spectra that will be needed. It is seldom a full analysis, and an analytical report based on interpretation of a single spectrum of a coating
should carry with it warnings about the kinds of materials that could
be obscured and the possible combinations of components that could
give rise to the spectrogram.

The prominent reference books on interpretation were described in Chap. 1 and a descriptive flow chart for interpreting spectra of polymers is provided in Chap. 12. Study of these data alone is not adequate for someone just beginning to do interpretation of spectra of unknowns. As minimum background training, a short course offered by universities, instrument companies, or consultants is recommended.

As a caution about interpreting spectra of mixtures, and to lend support for the repeated recommendations herein about fractionating samples before reporting spectral identifications, a simple interpretation situation is shown in Fig. 18. This example makes use of an easily interpretable portion of the spectrum: OH bands at 3260 cm^{-1}, and C-H stretching at 2960 to 2970 cm^{-1} for the CH in CH$_3$ and at 2940 cm^{-1} for CH in CH$_2$. The spectra clearly show the high methyl content of tertiary butanol and both the CH$_2$ and CH$_3$ bands in iso-butanol and n-butanol. An extrapolation of this relationship is made by the spectrum of n-octanol which has a considerably higher proportion of CH$_2$ groups to CH$_3$ groups. It shows that on a mole percent basis the concentration of OH is less in octanol than in the butanols. With these reference data it would be possible to interpret the spectrum of the "unknown" as a long-chain alcohol with a lesser concentration of OH than n-octanol. In fact, it is a mixed long-chain hydrocarbon, mineral oil, and a low-molecular weight alcohol, iso-butanol. This example demonstrates that functional group additivity occurs in a mixture from either multiple groups on one molecule or from different compounds in a mixture. Without several distinctive fingerprinting bands it is not possible to determine which functional group is on which molecular skeleton in a mixture of compounds. Complex coatings formulations frequently obscure the necessary fingerprinting bands, and fractionation becomes essential for a reliable analysis.

Moderately complex coatings often exhibit absorption bands at all of the well-known group frequency positions: OH, aliphatic CH, carbonyl, aromatic, C-H$_3$, C=O, etc. This situation is emphasized by Hummel and Scholl [3, p. 87] with a graph of the relative abundance

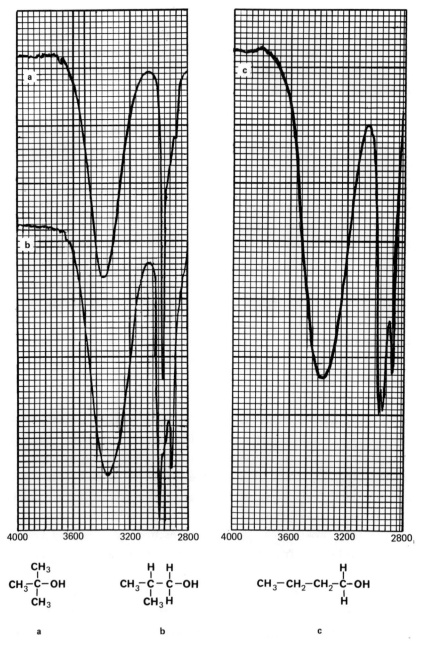

FIG. 18. Comparison of OH and CH_2 and CH_3 bands for alcohols (a-d) and a mixture of an aliphatic oil and an alcohol (e).

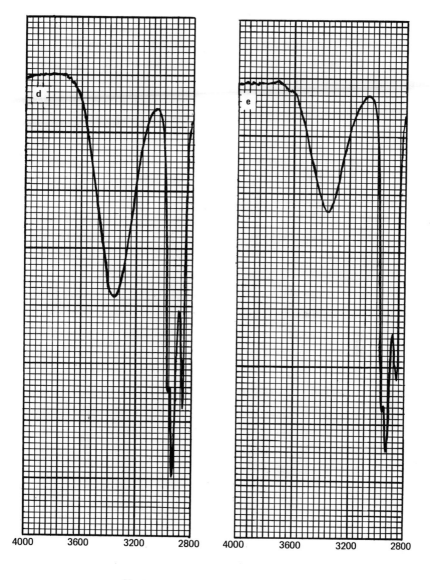

$$CH_3-(CH_2)_6-\overset{\displaystyle H}{\underset{\displaystyle H}{C}}-OH$$

d

"Unknown"

e

at each wavelength of bands for 2,000 spectra of polymers and gums. This relative abundance chart looks remarkably like the spectrum of many coatings samples. It shows that over 60% of these materials absorb at 3.4, 6.8, and 7.3 μm, and an average of over 30% absorb at every 0.1 μm between 8 μm and 8.6 μm. Thus most materials must have several absorption bands in addition to these to be identified completely.

Compounds with a repeatable pattern of strongly absorbing functional groups or functional groups that occur in a narrow, distinctive frequency range such as a triple bond can often be identified in a spectrum of a mixture. On the other hand, as is the case with the alcohol and alkane example above, it may not be possible for even a strong pattern of bands to be identified as being caused by one specific material.

The distinctive characteristics of phthalates, monosubstituted benzene rings, silicones, phosphates, and sulfonates are generally identifiable as compound classes in a mixture, but the nature of the molecule may still be misidentified: a monosubstituted aromatic can arise from polystyrene, styrene-butadiene rubber, styrene-modified resins, or retained toluene solvent; phthalate plasticizer in poly vinylchloride polymers has often been reported as a phthalic alkyd; aliphatic esters give very similar IR spectra regardless of whether they are from a varnish or a plasticizing oil.

It is recommended that spectra of these common polymers and additives which have strong absorption bands be obtained early in the operation of a coatings analytical laboratory and kept available for ready reference until they are memorized. A component which has strong, distinctive bands must be considered suspect as possibly masking other major constituents which exhibit weaker or broader bands and are less detectable in a mixture. (Refer to Figs. 1 and 5 for examples.)

If, following a first test, a spectrum of a coating is not recognizable or does not lend itself to the flow diagram treatment of Chap. 12 for spectral identification, there are two principal alternate choices for the next step:

1. Experiment with solvent clean-up. As simple a step as soaking the sample film which has already been run in a solvent may give distinct spectra of the soluble portion as well as the insoluble portion. That can lead to the identification of both fractions. This solvent test may be achieved without removing the sample from the plate.

2. Make a computer search of published reference spectra. Individual polymers are easily identified by a computer search. A commercial product whose spectrum has been entered into the data bank as a mixture can also be retrieved readily. The chance for identification is obviously enhanced as compounds are purified. However, there is one system available internationally in time sharing to accommodate impure samples [27].

Final spectral interpretation for positive identification of a compound should be based on point-by-point comparison with a reference spectrum of a known material. Ideally a reference spectrum is run on the same spectrometer under the same operating conditions as the sample. If complete assurance is needed that no additional components remain undetected in the spectrum, a quantitative standard should be prepared.

The most preferred quantitative procedure is comparison of solutions of known concentration in a fixed thickness cell. The preferred IR transparent solvents (CS_2, CCl_4, $CHCl_3$) do not dissolve many coatings materials, but other solvents with limited regions of transparency can be used. If solutions are impossible, quantitative preparation in a KBr disk or preparation by weight with an internal standard in a mull can be accurate to a few percent, if adequate experimental caution is taken. For techniques refer to Chap. 1 and to Potts [21] and Smith [28].

Exact band positions, relative band intensities, and band shapes should all be evaluated carefully when completing spectral identification. If a band in the unknown appears to be of a slightly different shape than the known, 30% or more of the sample may still be unidentified. A difference spectrum with the known component in the

reference beam is very effective in detecting additional components
hidden by overlapping bands. Considerable care must be used with
this technique [28, pp. 3712-3717] but it is an invaluable interpretation aid when properly carried out.

b. *Raman*. The interpretation of Raman spectra of organic
materials is discussed in the chapter by Nyquist and Kagel (Chap. 6).
IR and Raman spectra are complementary in many ways for diagnostic
interpretation of chemical structures. One of the structures for
which IR spectra have been most inadequate is the olefinic C=C in
oils and oil-based coatings. O'Neill and Falla [29] show comparative
spectra of IR and Raman spectra of linseed oil, tung oil, and pentaerythritol linseed phthalic alkyd resins in which the strongest bands
in the Raman spectra are the C=C stretching bands. They are weak in
the IR spectra. Aromatic ring substitution, α-substituted nitrile
groups, and the distinction between $-NH_2$ and -OH groups are among the
diagnostic bands for which Raman spectra may be used in conjunction
with IR spectra to provide more complete structural analysis than
can be obtained by either method alone [17].

B. The Analysis of Formulated Coatings

Identification of some of the components or the type of coating
from a spectrum of the total sample and preliminary solvent separation
tests permit development of an analytical scheme. This scheme ideally
includes the use of weighed samples followed by quantitative physical
separation of individual constituents. If these isolated components
are esters or copolymers, chemical separation of the reactants may
also be necessary.

It is advantageous to separate plasticizer and other additives
from the bulk of the sample. If the additive is a weak absorber it
is necessary to separate it in order to identify it; if it is a strong
absorber, even at 1 or 2% in concentration it may have bands as strong
as bands of some polymers. Treatment of finely divided samples with
a light hydrocarbon such as n-heptane dissolves most plasticizers.

Exhaustive extraction as provided by Soxhlet equipment is an easy
way to effect the separation. For some additives carbon tetrachlor-
ide, tetrachloroethylene, alcohol, or even water may also be useful
extractants. Benzene or toluene are powerful solvents for many additives, but they also dissolve some polymers. Low-molecular-weight
portions of polymers are likely contaminants of plasticizer or additive fractions if the solvent used is too powerful. Preliminary
tests are necessary to establish the optimum solvent choices for any
particular coatings system.

The plasticizer extraction is followed by solvent separation of
the binders of film formers. Aromatic hydrocarbons, acetone, methyl
ethyl ketone, o-dichlorobenzene, and dimethylformamide are common
solvents for polymers and resins.

Some polymers are particularly tenacious in the retention of
plasticizer. Complete separation may not be possible without dis-
solving the polymer and then reprecipitating it by pouring the solution into an incompatible solvent. A lesser amount of third component (water or alcohol, for example) may be helpful in precipitating
the polymer or providing solvent phase separation.

The extracted additives may be separated into polar and nonpolar
components by their distribution between immiscible solvents. Chro-
matographic separation may be necessary before pure enough fractions
are obtained for complete identification.

Once the polymer fraction is free from additives all of the
ordinary polymer identification procedures can be employed. For some
analytical requests, for example in establishing or protecting a
patent, a particular molecular weight range or copolymer ratio may
be important. For these samples, gel permeation followed by spectral
characterization of the fractions may be necessary. Nuclear magnetic
resonance data may be required to distinguish between block copolymers
and random copolymers.

Coatings commonly contain a wide assortment of pigments, dyes,
fillers, preservatives, antioxidants, oxidation catalysts, etc. For
many analytical purposes the important point about these materials
is to keep them from interfering with the analysis of the rest of

the sample, i.e., most analytical requests do not require their
complete identification. On the other hand, without determining
what the additives are, or at least obtaining spectra for comparison
purposes, the analyst may be led into a misinterpretation of the
spectrum for the principal components. Obtaining spectra and the
weights of all separated fractions cannot be too strongly recommended for even a semiquantitative analytical procedure.

Typical case histories of two analyses--one transmission, one
internal reflection--follow.

Case History: Analysis of Coating on Transfer Paper. Transfer paper
analysis is usually desired for monitoring of patent infringements or
performance failure, and therefore a quantitative analysis of the
major binders and plasticizers is required.

The coating can be scraped from the paper surface with a razor
blade (Fig. 11) and a sample of 1 g or more is generally obtained
from a single sheet. This is a convenient size sample because components as low as 1% give a large enough fraction to permit quantitative sample transfer and easy sample handling for spectral determination. Extraction of the sample with nC_7 removes the plasticizer, and
extraction with benzene followed by methyl ethyl ketone dissolves the
polymers. The major component of the insoluble residue from these
extractions is usually carbon. Patents seldom are concerned with
this portion of the sample. However digestion with other powerful
solvents such as dimethylformamide or o-dichlorobenzene, or pyrolysis
(to eliminate the possibility of additional polymers being present),
elemental analysis, or ash determination may be necessary for con-
firmation of its identity.

The plasticizer fraction is often about 40% of the sample and
may even be a mixture. Separation of the plasticizer components, by
immiscible solvent systems or by chromatography, generally permits
complete analysis.

The composition of the binder is often the patented composition
in a transfer coating. Analytical assurance of the presence of even
a few percent modification can be very important. Solvent reprecip-
itation and subsequent extraction may be necessary to obtain an

uncontaminated specimen. A particular problem exists with strongly
absorbing dyes of the crystal violet family. For example, a very
small amount of crystal violet can give a spectral contribution
equivalent to several percent ethylacrylate in polymethyl methacrylate
at the wavelength where ethylacrylate is detectable. See Fig. 7 for
the small spectral difference involved in this particular acrylic
resin modification.

The use of known blends to determine the limits of sensitivity
of the combined fractionation and spectral determination procedure
is fully as important for spectroscopic analysis as it is for chemical analysis.

Spectra of plasticizer and binder fractions from a transfer
paper are presented in Fig. 19. Also included for comparison is an
internal reflection spectrum of the total sample. This coating is
a copolymer plasticized with a glycerol ester of long-chain fatty
acids with a crystal violet dye and carbon filler. The spectrum of
the oil (heptane extract) is curve (a). Curve (b) is the film former,
polyvinyl acetate - polyvinyl chloride copolymer. The band at 1580
cm^{-1} is not attributable to the plasticizer or to the polymer. It
is caused by the dye.

These components are evident but not identifiable with certainty
in spectrum (c), which is the internal reflection spectrum of the
total specimen. An internal reflection spectrum requires less than
an hour because no sample separation is involved. It does not provide an analysis. Fractionation and spectral determination complete
with quantitative analysis require several hours. To obtain complete
analytical details for some samples, several days may be necessary.

A good use of the internal reflection technique for this kind
of sample is for a preliminary screening to establish the likely
kinds of polymers present as an aid in making solvent selections for
the fractionation. High carbon content usually interferes less with
internal reflection spectra than with transmission spectra.

Case History: Analysis of Adhesive Tape. The complexity of construction of adhesive tape is shown in Figure 20. An analysis of the
total coating does not establish which component is in any given layer.

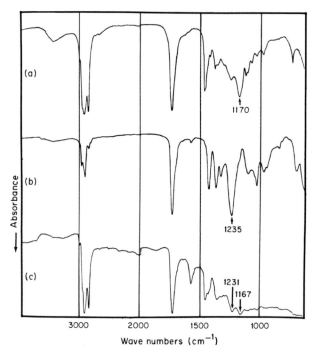

FIG. 19.   IR spectra of a coating from transfer paper: (a) the ester
oil plasticizer extracted with n-heptane; (b) film-former isolated
from coating and identifiable as polyvinyl chloride - polyvinyl acetate copolymer (refer to Fig. 6 for comparison); (c) internal reflection spectrum of total sample before fractionation.

This information can be best obtained by internal reflection spectra
of each separable stratum.   These spectra suggest the fractionation
procedures to be followed and permit reconstruction of the adhesive
tape system once the backing film and components of each coating
layer are identified.

An example of spectrometric analysis of a tape is given in Fig.
21.   Spectrum (a) of the tape adhesive surface shows isoprene and a
small amount of probable ester.   In addition a band suggestive of
carboxylic acid salt at 1580 cm^{-1} and possible strong long-wavelength
absorption are evident only in this layer.   Thus when zinc resinate,
zinc oxide, and titanium dioxide are isolated subsequently by fractionation procedures, it is possible to ascribe them to the surface
layer of the adhesive.

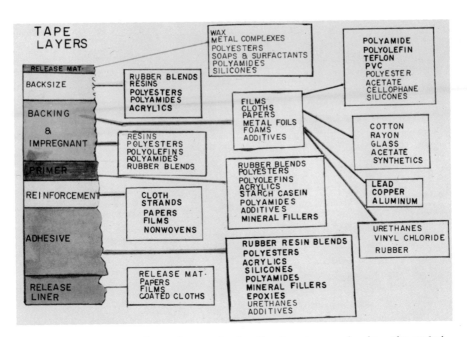

FIG. 20. Construction of a typical adhesive tape, showing the multi-plicity of probable components. Reprinted from Ref. 25, courtesy of C. Paralusz and J. Luongo.

Paralusz [25] describes an interesting transfer technique for separating the adhesive layer from the tape backing to expose the interface layers. A strip of Mylar is placed against the adhesive surface and then peeled away by a stripping motion, which peels the Mylar® back 180° with the bulk of the adhesive transferred to it. This technique permits obtaining spectra (b) and (c) in Fig. 21. After butadiene-acrylonitrile copolymer is identified positively by fractionation, its position in the tape strata can be assigned to the primer surface through its distinctive band in curve (c).

The transmission spectrum, (d), of the film backing permits its identification as polyvinyl chloride with carbonyl component(s) as either copolymers or plasticizers. Subsequent extraction completes the identification.

Use of difference spectroscopy, with the coated sample in the sample beam and the film with the coating removed in the reference beam, effectively cancels polyvinyl chloride from the spectrum

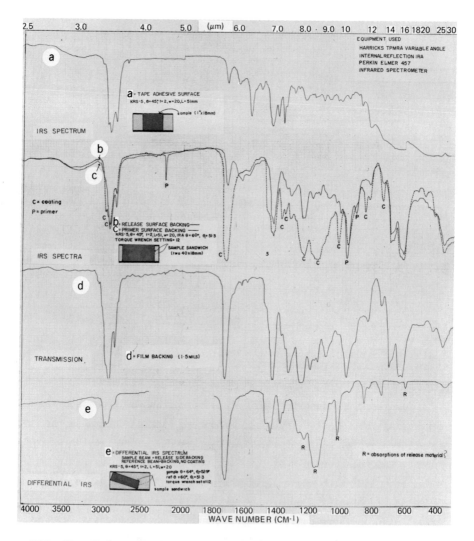

FIG. 21.   Infrared spectra obtained on layers of an adhesive tape.
See text for spectral interpretation.   Reprinted from Ref. 25,
courtesy of C. Paralusz.

[curve (e)] and permits the suggestion that the outer coating is an
ethylacrylate polymer with other components present.   A direct analysis of the removed coating might frequently be preferable to this use
of difference spectroscopy.

Difference spectra can be extremely effective in both internal reflection and transmission spectroscopy for comparison of closely similar materials or for in situ analyses described in the following sections.

C. Comparison of Closely Similar Materials

Crude oil, asphalt, coal, and cellulose are important commercial examples of materials whose spectra have broad regions of strong absorption from overlapping bands. To distinguish between crude oils from different producing horizons or different geographical regions requires a comparison of small spectral differences superimposed on many spectral similarities. The same is true for spectra of coal or for paper and starch products.

To obtain useful data from spectra of such materials, the suppressed zero method in single-beam operation was developed in 1946 [30]. This work showed that if the spectrum of a sample prepared so thick that it transmits only a few percent of the incident radiation is run with wide spectrometer slits so that the 100% line is manyfold off scale (i.e., with zero suppressed or the scale expanded), small percentage differences between the samples give a characteristic absorption pattern detectable above the general high level of absorption. Extensive use of this procedure has been made to relate crude oil-producing zones across a wide geographical area or through geological strata, and to establish the origin of oil seeps and tar sands [31].

Double-beam spectrometers permit a still further enriched spectrum from difference spectra by eliminating from the recorded data all of the common composition characteristics. For example, asphalts made from a wide range of crude oils provide the grossly similar spectra shown in Fig. 22. The major bands are predictably all the same and the identifying structures relate to the proportion of aromaticity and aliphatic character and to differences in the kinds of aromatics present. Carbonyl absorptions are prominent but relate

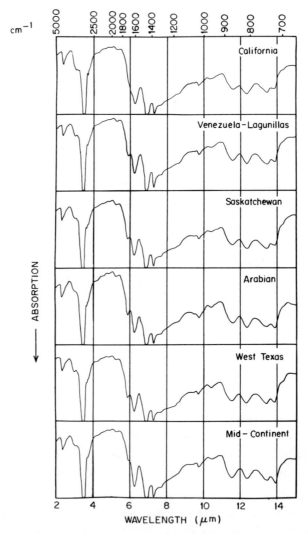

FIG. 22.   IR absorption spectra of asphalts from six crude oil sources.
Reprinted from Ref. 32, p. 155, by courtesy of the American Chemical
Society.

principally to the processing history of the asphalt rather than to
the hydrocarbon composition which is indicative of the origin and
nature of the sample.   Difference spectra obtained with one of the
asphalts which represents a composition extreme was used to record

the spectral differences between it and other asphalts. That procedure permits characterization of the chemically significant features
which relate one asphalt to another in geological origin and composition (Fig. 23).

It has been demonstrated that sources of asphalts can be identified by comparison of the amount of $C-CH_3$ and $(CH_2)_x$ chains that are
present as a function of the total C-H absorptivity at 2940 cm^{-1} [32].
In terms of practical applications it is possible to predict the
weatherability of roofing asphalts on the basis of the amount and type
of CH_2 chain present as shown by the 720-cm^{-1} band (Fig. 24).

In the course of this work it was learned that broad boiling
petroleum fractions are similarly useful for comparing sources of
crude oils. In fact, gasoline seeping into the street or neighboring
basements from a leaking, buried storage tank can be quickly traced
by attention to the relative intensities of the many absorption bands
in the IR spectrum. It is sometimes necessary to discount bands attributable to the most volatile components in such spectral comparisons.

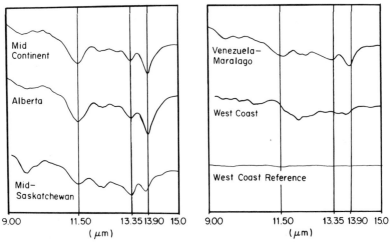

FIG. 23. IR difference spectra of roofing grade asphalts show great
similarity between asphalts from mid-continent and Alberta crude oils.
Distinctive characteristics are observed for each major asphalt type.
Minor differences are detectable between west coast asphalts. Data
from 10% solutions in CS_2 in 1-mm cell. Reprinted from Ref. 32, p.
156, by courtesy of the American Chemical Society.

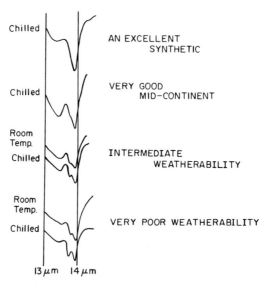

FIG. 24.   IR absorption spectra showing paraffinic crystallinity (13.7- and 13.9-μm doublet) in poor-weathering coating asphalts. Reprinted from Ref. 32, p. 159, by courtesy of the American Chemical Society.

The bands which are most important are the ones which are indigenous to the nature of the production horizon and refinery processing rather than the ones that relate to contamination and oxidation. For thin films coating water or rock it is especially important that the hydrocarbon distribution be considered in order to identify the source of an oil layer.

In all identifications relating to source of an oil spill or a gasoline leak, it must be remembered that one source of crude oil and one refinery product may be handled by different shipping lines or different distributors. Therefore, only if spectral differences are observed for known samples from different suspected polluting sources can it be ascertained which one is the origin of a given sample; that is, analytical data of any kind should be used only to select the best match among samples, not to provide the conclusion that a given sample arises from only one possible source.

IR spectra of celluloses are closely similar, yet spectral differences can be observed for mature wood and for freshly formed wood
[33] and between papers made from different proportions of kraft and
groundwood stock [34]. The differences which relate to small amounts
of lignin-like structures in groundwood are measured quantitatively
to a few percent by differential transmission spectra (Fig. 25) and
are evident for greater concentration differences in internal reflection spectra (Fig. 26).

D. In situ Analysis
    of Surface Composition

    In situ measurement of spectra of coatings is practiced in a
wide range of applications:

    1. Transparent coatings for cans by double transmission with
reflection from the can surface or by internal reflection

    2. Wire coatings by internal reflection [35,36]

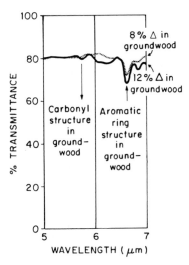

FIG. 25. Transmission difference spectra of paper samples with high-
er groundwood furnish concentration in the paper in the sample beam.
Reprinted from Ref. 34, p. 1700, by courtesy of the American Chemical
Society.

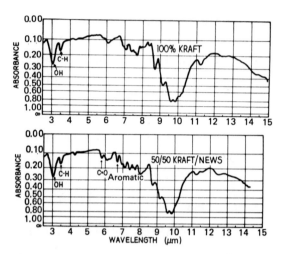

FIG. 26. Multiple internal reflection spectra of paper show differences in lignin content between furnishes of 100% Kraft and 50/50 Kraft/News. Reprinted from Ref. 34, p. 1701, by courtesy of the American Chemical Society.

3.   Drying oil reactions at surfaces by internal reflection [37]

4.   Chemical modification of cotton fibers by transmission through KBr disks or by multiple internal reflectance [38,39]

5.   Pigmented coated papers by internal reflection spectroscopy [40]

6.   Urea resin concentration on paper fibers by transmission [41]

7.   Curing and degradation reactions of any coatings systems for which there is a change in chemical functionality as drying, curing, or degradation takes place.

The key techniques necessary for in situ spectroscopy are the same as those described by Crawford and Swanson in Chap. 1. It is important that instrument operation be optimum because many of the systems such as fibers, glass wool, paper sheets, and pigmented coatings scatter the incident radiation and others such as asphalt, crude oils, coal, or carbon black absorb strongly. Both of these situations lead to reduction in energy of the system, and good results depend upon optimizing the operating conditions of the spectrometer. Fortunately, slits may be opened considerably beyond the standard spectrometer settings for most of the bands in polymeric samples, without adversely affecting peak discrimination; that is, most bands in polymers are broad and not very sensitive to resolution. A few bands, e.g., nitrile groups, can be significantly affected by resolution, however, so it cannot be assumed that the optimum instrumental compromises for one kind of sample are also optimum for another.

Scattered light can be reduced considerably by the use of an immersion oil closely matching the refraction index of the sample [33,34].

In situ measurements lend themselves to sample misidentification because of spectral obscuration as noted in the earlier sections. Measurements in situ are most effective for studying chemical structure changes in a known material or for quantitative analysis of systems whose qualitative composition is known.

Development of analytical spectroscopic procedures has been widely practiced by the instrument manufacturing companies, and bulletins on both qualitative and quantitative coatings analyses are available from Perkin-Elmer Corporation, Beckman Instruments, Cary Instruments, Wilks Scientific, and Barnes Engineering. These procedures are valuable because of the high level of expertise and industrial experience of the authors, and they are available free of charge upon request.

IV. RESEARCH ON INTERFACES

Answering questions relative to the nature of a surface, the orientation of molecules at an interface, or the nature of chemical-physical bonding between surfaces relies on quite different techniques and experimental approaches than conducting an analysis to identify the components of a coating.

The reasons for wanting to do fundamental research on interfaces cover a large range of significant possible areas. The research may be basic in terms of the nature of the chemical structures involved, but a far-reaching application is to provide a basis for understanding

the nature of adhesive or nonadhesive surfaces, chemically resistant
or susceptible surfaces, catalytically active or reaction-poisoning
surfaces, or chemical reactions at electrodes. Practical questions
about surfaces range from how to make an automobile paint maintain
adhesion under impact or how to speed fabrication of metal or plastic
products, to what controls the adhesion of body tissues to prosthetic
devices. What is it that holds cells together? And conversely, what
is it that causes cells to separate in metastasis or what will prevent a plastic artery from plugging?

A. Experimental Considerations

Although spectroscopy is the most powerful tool available, it
has been severely limited in these various applications because it
can be expected that the most critical bond lengths are only a few
angstroms, whereas one-pass transmission spectra require pathlengths
in the range of 0.02 mm and even internal reflection spectroscopy
generally involves sample penetrations of the order of a micrometer.
Experiment design has been dictated largely by energy requirements
to actuate a recorder or to make a photographic image. Data digi-
tizing and storage by computers together with computer averaging
techniques are changing this restrictive dependence upon detected
energy. Thus, these techniques along with Fourier transform spectroscopy and its capability of greater spectral through-put and simul-
taneous determination of many spectral frequencies, is permitting
more fruitful research into orientation of molecules at a surface.

In order to build up the intensity of an absorption band from
a very thin sample layer or from an interface between surfaces, it
is necessary to cause the radiant energy to interact repeatedly with
a small sample thickness. This interaction can be achieved by
multiple transmission, by multiple internal reflectance, or by en-
riching the concentration of surface layers in a bulk sample by preparing it as a fine powder prior to treatment with the interacting
compound.

Transmission spectra of a thin film can be obtained if it is coated on reflective surfaces and the beam is propagated at an angle between the reflective plates. This technique is limited by the requirement that one surface of the two-phase system be reflective. Front surface reflection is superimposed on the double-transmission curve, and the nature of the substrate (metallic or dielectric) also affects the observed spectrum. The result is a complex set of variables involved in the interpretation of experimental data for very thin films.

Internal reflection spectra, as already discussed, provide a way of obtaining spectral data on the surface layer of a material with minimum penetration into the sample bulk. Thus, the absorbance of atomic groups near the surface is relatively richer in internal reflection spectra than in transmission spectra.

In both transmission and internal reflection spectroscopy of thin films, the physics of light propagation and the properties of electromagnetic radiation reflected from a surface play considerably larger roles in data interpretation than need to be considered for practical chemical purposes in interpreting spectra of bulk samples [23]. Development of this subject has moved rapidly because of its importance in solid state devices and optoelectronics. A discussion of light waves in thin films [42] and detailed consideration of the variables applicable in both transmission and internal reflection spectra [43] should be helpful to anyone planning to do research involving absorption intensities at interfaces.

Despite the limitations described thus far, IR spectroscopy and, now to an increasing degree, Raman spectroscopy are two of the most powerful tools available to surface chemists investigating chemical bonding.

B. Reported Results

Hair [44] has written a book which provides readily understood theory on the adsorption process and the nature of the surface bond. He details the successes and experimental disadvantages of the many techniques that have been tried in the past. The most dramatic investigations have been those in which a significant difference in the chemical composition of a compound occurs when it is adsorbed. The investigation of Si-OH groups on the surface of silica has included reaction of the OH with trimethylchlorosilane and subsequent deuteration of remaining OH groups [45]. The reaction between $SiCl_4$ and surface hydroxyl groups has been demonstrated to involve two hydroxyl groups per $SiCl_4$ forming -O, Cl groups and has been used

$$-O \diagdown \underset{Si}{\diagup} Cl$$
$$-O \diagup \diagdown Cl$$

to investigate the changes that a freshly prepared gel surface undergoes during aging [46].

Surface fluorination, surface esterification, surface reactions with $CHCl_3$, hydrocarbons, HCN, and diborane are all reported on and the degree of uniformity in observations among different investigators are discussed. Hair cites 84 references to studies on the silica surface alone, and has additional chapters on investigations of adsorption on alumina, zeolites, clay, other oxides, carbonates, and metals.

Adsorption studies of carbon monoxide on various metal substrates resulted in calculation of force constants for platinum-to-carbon and carbon-to-oxygen bonds which demonstrate essentially single-bond character for the former and triple-bond character for the latter (both with some probable double-bond character) [47]. Subsequent similar studies on iron, nickel, copper, ruthenium, and palladium are well described and referenced by Hair [44].

Laser Raman spectra of adsorbed species are being investigated to gain an understanding of the nature of surface-adsorbate interactions and to distinguish between electron acceptor and proton donor sites on silica gel surfaces [48,49]. Both pyridine and piperidine adsorbed at low surface coverage are found to be hydrogen-bonded, probably to hydroxyl groups on the silica gel surface. On the other hand, 2-chloropyridine shows no spectral difference from the liquid form in an equivalent experiment.

The advantage of Raman spectroscopy over IR spectroscopy in the study of adsorbed species on metal oxides stems from the fact that IR bands of the metal-oxygen stretching vibration are strong and often interfere with bands of the adsorbed species. Raman scattering, on the other hand, shows weak bands for metal oxides. Moreover, the entire spectral range of 100 to 4000 cm^{-1} can be investigated.

Considerable documentation of IR spectral band frequency and intensity changes for pyridine with hydrogen bonding shows the 1400- to 1600-cm^{-1} region as the most useful [50]. In addition, Hendra et al. have investigated details of the Raman spectrum of pyridine for the 991- and 1031-cm^{-1} bands, which are the most prominent ones in the Raman spectra of adsorbed pyridine, and report shifts of the 991-cm^{-1} band to be particularly diagnostic for the environment of pyridine. Hendra et al. [51] describe experimental details and give data for distinguishing between chemisorption and physisorption for a variety of metal oxide surfaces (see Table 4). For some of the substrates investigated by Hendra, marked differences in the nature of pyridine adsorption were shown when the concentration of pyridine was low, e.g., 93 $Å^2$/molecule for silica gel (see Table 5).

TABLE 4

Details of the Raman Spectra of Pyridine
onto Oxide Surfaces at Low Coverages[a]

Oxide substrate	Wave number of pyridine bands at minimum coverage	Interpretation
Magnesium oxide	991	No bonding to surface (i.e. liquid pyridine)
NH_4^+/Mordenite (at monolayer coverage)	1004	H-bonding of pyridine to the surface
Silica gel (at 93 $Å^2$/molecule)	1007	Pyridine strongly bound to -OH sites
Titanium dioxide	1016	Coordinately bound (Lewis) pyridine

[a]Adapted from Ref. 51, p. 1767, by courtesy of J. Chem. Soc.

TABLE 5

Details of the Raman Spectrum of Pyridine in a Range of
Environments in the Spectral Range from 990 to 1100 cm^{-1} [a]

System	Wave number	(intensity)			
Pyridine (liquid)	991	(10)		1031	(8)
Pyridine-chloroform	993	(10)		1032	(7-5)
Pyridine-ethanol	999	(10)		1032	(6-5)
Pyridine-water	1003	(10)		1036	(5)
Pyridinium chloride (soln.)	1010	(10)	1028 (3)	1060	(1)
C_5H_5N-O (solid)	1016	(10)		1043	(1-5)
C_5H_5N-O$\cdot$2H$_2$O (solid)	1018	(10)		1043	(1-5)

[a]Abridged from Ref. 51, p. 1767, courtesy of J. Chem. Soc.

Myers [52] postulated a relationship between vibration frequencies and catalytic reactions, and Gardner mathematically interpreted the relationship between catalytic activity of different metals
in the oxidation of CO to CO_2 and the frequency of CO chemisorption
on the metals [53].

A versatile high-temperature, high-pressure cell for in situ
catalyst investigation which allows reaction conditions of up to
100 atmospheres at 200°C has been described by Tinker and Morris [54].
Thus, considerable and fruitful research has been carried out over
the past 20 years on spectroscopic investigation of chemisorbed
species and the relationship of spectral data to reaction mechanisms.

It is possible to observe more subtle spectral differences than
those caused by chemical bonding in systems which are comparatively
chemically inert.

In a classic investigation, McDonald [55] studied the effect of
introducing nonpolar gases onto the surface of silica at -190°C. He
demonstrated that in all cases the OH stretching band was lowered in
frequency as gas was adsorbed on the surface in increasing amounts.
It was possible to interpret the variations in response to different
adsorbed gases as showing a preferential interaction of the surface
OH with nitrogen, as compared to much weaker interaction with oxygen
or argon.

Experimental work using internal reflection and polarization techniques with a Fourier transform spectrometer is being conducted at Battelle Memorial Institute [56] to demonstrate molecular order of thin film lubricants and film-to-surface bonding. From this work it is indicated that in order for the small change in molecular order at an interface to be characterized, digitized data and storage of band ratio pairs of polarized data are required. Considerable experimental caution is necessary to discount effects of anomalous polarization and then to detect polarization differences between spectra of interface surfaces and bulk liquids.

Biological interfaces are among the most complex systems being systematically investigated. Research aimed at determining the surface characteristics necessary for candidate biomedical materials to be as nonthrombogenic as possible has required a combination of techniques: electron microscopy [57], contact angle measurements on biological films adhering to polymers, and multiple internal reflection measurements of the film interfaces [58,59]. Spectra of polypeptide monolayers reveal molecular orientation at interfaces by both IR [59] and Raman spectroscopic techniques [60].

REFERENCES

1. Infrared and Spectroscopy Committee of the Chicago Society for Paint Technology, (L. C. Afremow, Chrmn.), and DeSoto, Inc., *Infrared Spectroscopy: Its Use in the Coatings Industry,* Federation of Societies for Paint Technology, Philadelphia, Pa.. 1969.

2. C. D. Smith-Craver, in *Characterization of Coatings: Physical Techniques (Treatise on Coatings, Vol. 2),* Part I (R. R. Myers and J. S. Long, eds), Marcel Dekker, New York, 1969, pp. 429-500.

3. D. O. Hummel and F. Scholl, *Infrared Analysis of Polymers, Resins and Additives: An Atlas,* Vols. 1 and 2, Wiley-Interscience, New York, 1968.

4. J. Haslam and H. A. Willis, *Identification and Analysis of Plastics,* Van Nostrand, Princeton, N.J. 1965.

5. C. P. A. Kappelmeier, *Chemical Analysis of Resin-Based Coating Materials,* Wiley-Interscience, New York, 1959.

6. American Society for Testing Materials, *Annual Book of ASTM Standards,* Philadelphia, Pa.

7. ASTM Committee E-13 (R. T. O'Connor, Chrmn.), *Manual on Recom-
    mended Practices in Spectrophotometry,* American Society for
    Testing Materials, Philadelphia, Pa., 1966.

8. *Report of ASTM Committee E-13,* American Society for Testing
    Materials, Philadelphia, Pa., March 1974.

9. G. G. Sward and H. A. Gardner (eds.), *Paint Testing Manual:
    Physical and Chemical Examination of Paints, Varnishes, Lacquers,
    and Colors,* 13th ed., American Society for Testing Materials,
    Philadelphia, Pa., 1972.

10. M. H. Swann, in *Paint Testing Manual: Physical and Chemical
    Examination of Paints, Varnishes, Lacquers, and Colors,* 13th
    ed. (G. G. Sward and H. A. Gradner, eds.), American Society for
    Testing Materials, Philadelphia, Pa., 1972, pp. 92-118.

11. *Modern Plastics Encyclopedia,* McGraw-Hill, New York, annual.

12. R. R. Myers and J. S. Long (eds.), *Treatise on Coatings: Film-
    Forming Compositions,* Part I, Marcel Dekker, New York, 1967.

13. Federation of Societies for Paint Technology, *Federation Series
    on Coatings Technology,* 3rd ed., Federation of Societies for
    Paint Technology, Philadelphia, Pa., 1972.

14. *Coatings and Plastics Preprints,* preprints of papers presented
    at National American Chemical Society Meetings, published by
    the Division of Organic Coatings and Plastics Chemistry,
    American Chemical Society, Washington, D.C., (annual).

15. C. L. Mantell, in *Film-Forming Compositions (Treatise on Coatings,
    Vol. 1),* Part I, (R. R. Myers and J. S. Long, eds.), Marcel
    Dekker, New York, 1967, pp. 341-390.

16. J. D. McGinness, in *Paint Testing Manual: Chemical Examination
    of Paints, Varnishes, Lacquers, and Colors,* 13th ed. (G. G.
    Sward and H. A. Gardner, eds.), American Society for Testing
    Materials, Philadelphia, Pa., 1972, pp. 495-499.

17. H. J. Sloane, in *Polymer Characterization: Interdisciplinary
    Approaches* (C. D. Craver, ed.), Plenum Press, New York, 1971,
    pp. 15-36.

18. O. F. Folmer, Jr., in *Polymer Characterization: Interdisciplin-
    ary Approaches* (C. D. Craver, ed.), Plenum Press, New York,
    1971, pp. 231-248.

19. C. E. R. Jones and G. E. J. Reynolds, *J. Gas Chromatog., 5,*
    25 (1967).

20. J. K. Haken, in *Characterization of Coatings: Physical Tech-
    niques (Treatise on Coatings, Vol. 2),* Part I (R. R. Myers and
    J. S. Long, eds.), Marcel Dekker, New York, 1969, pp. 249-262.

21. W. J. Potts, Jr., *Chemical Infrared Spectroscopy,* Vol. 1, Wiley,
    New York, 1962.

22. The Coblentz Society, Inc., an association of persons interested in fostering the understanding and application of infrared spectroscopy and related fields, 761 Main Ave., Norwalk, Conn. 06851.

23. N. J. Harrick, *Internal Reflection Spectroscopy,* Wiley-Interscience, New York, 1967.

24. J. W. Cassels, A. C. Gilby, and P. A. Wilks, Jr., *Appl. Spectrosc.,* *24*(5), 539 (1970).

25. C. Paralusz, *J. Colloid Interface Sci., 47, 3* (1974).

26. G. E. McGraw, in *Polymer Characterization: Interdisciplinary Approaches* (C. D. Craver, ed.), Plenum Press, New York, 1971, p. 37-46.

27. C. D. Craver and E. Singer, Society of Applied Spectroscopy, 10th National Meeting, St. Louis, Mo., October 1971.

28. A. L. Smith, in *Treatise on Analytical Chemistry,* Part I, Vol. 6 (I. M. Kolthoff and P. Elving, eds.), Wiley-Interscience, New York, 1965.

29. L. A. O'Neill and N. A. R. Falla, *Chem. Ind.,* Nov. 20, 1971, pp. 1349-1351.

30. C. D. Smith-Craver, U.S. Patent 2,648,010, 1946.

31. J. F. Black and C. D. Smith-Craver, U.S. Patent 2,700,593, 1947.

32. C. D. Smith-Craver, C. C. Schuetz, and R. S. Hodgson, *Ind. Eng. Chem. Prod. Res. Dev., 5,* 153 (1966).

33. R. G. Zhbankov, *Infrared Spectra of Cellulose and Its Derivatives,* Plenum Publishing Corp., New York, 1966.

34. C. D. Smith-Craver and J. K. Wise, *Anal. Chem., 39,* 1968 (1967).

35. C. F. Hunt and A. H. Markhart, *Insulation,* Lake Publishing, Libertyville, Ill., 1960.

36. R. J. McGowan, *Anal. Chem., 41,* 2074 (1969).

37. A. E. Rheineck, R. H. Peterson, and G. M. Sastry, *J. Paint Technol., 39* (511), 484 (1967).

38. R. T. O'Connor, E. F. DuPre, and E. R. McCall, *Anal. Chem., 29,* 998 (1957).

39. E. R. McCall, S. H. Miles, and R. T. O'Connor, *Amer. Dyestuff Rep., 55*(11), 31 (1966).

40. J. P. Deley, R. J. Gigi, and A. J. Liotti, *Tappi, 49*(8), 57A (1966).

41. J. K. Wise and C. D. Smith-Craver, *Anal. Chem., 39,* 1702 (1967).

42. P. K. Tien, *Applied Optics, 10* (11), 2395 (1971).

43. W. W. Wendlandt and H. G. Hecht, *Reflectance Spectroscopy,* Wiley-Interscience, New York, 1966.

44. M. L. Hair, *Infrared Spectroscopy in Surface Chemistry*, Marcel Dekker, New York, 1967, p. 45.

45. G. A. Galkin, S. Zhandov, A. V. Kiselev, and V. Lygin, *Kolloidn. Zh., 25* (1), 123 (1963); and *26* (3), 324 (1964).

46. J. B. Peri, *J. Phys. Chem., 70, 2937* (1966).

47. R. Eischens, S. A. Francis, and W. A. Pliskin, *J. Phys. Chem., 60*, 194 (1956).

48. N. T. Tam, R. P. Cooney, and G. Curthoys, *Raman Newsletter, 48,* 10 (1972).

49. R. O. Kagel, *J. Phys. Chem., 74,* 4518 (1970).

50. E. P. Parry, *J. Catal., 2, 371* (1963).

51. P. J. Hendra, J. R. Horder, and E. J. Loader, *J. Chem. Soc., (A),* 1766 (1971).

52. R. R. Myers, *Ann., N. Y. Acad. Sci., 72,* 341 (1958).

53. R. A. Gardner, *J. Catal., 3,* 7 (1964).

54. H. B. Tinker and D. E. Morris, in *Rev. Sci. Instr., 43* (7), (1972).

55. R. S. McDonald, *J. Amer. Chem. Soc., 79,* 850 (1957).

56. R. Jakobsen, Pittsburgh Conference on Analytical Chemistry and Applied Spectroscopy, Cleveland, Ohio, March 1974.

57. R. C. Dutton, A. J. Webber, S. A. Johnson, and R. E. Baier, in *J. Biomed. Mater. Res., 3,* 13 (1969).

58. R. E. Baier, R. C. Dutton, and V. L. Gott, in *Surface Chemistry of Biological Systems,* Plenum Press, New York, 1970.

59. R. E. Baier and G. I. Loeb, in *Polymer Characterization Interdisciplinary Approaches,* (C. D. Smith-Craver, ed.), Plenum Press, New York, 1971, p. 79-96.

60. W. L. Peticolas, E. W. Small, and B. Fanconi, in *Polymer Characterization Interdisciplinary Approaches,* (C. D. Smith-Craver, ed.), Plenum Press, New York, 1971, p. 47-77.

AUTHOR INDEX

Part A, pages 1-346; Part B, pages 347-716; Part C, pages 717-1000

Numbers in brackets are reference numbers and indicate that an author's work is referred to although his name is not cited in the text. Italicized numbers give the page on which the complete reference is cited.

A

Abe, M., 764[126,128], 772[126, 128], 773[126,128], 774 [128], *866*
Abe, Y., 744[49,50], 746[50], 750[49], *863*
Abel, E. W., 225, *275*
Abrahams, D. H., 712[46], *715*
Abramson, B., 848[225], *871*
Achhammer, B. G., 705[7], *713*
Adam, H. K., 857[236], 858[236], *871*
Adams, D. M., 108[8], 109[9], 168[88,90], *202, 205,* 248 [54], *276*
Agron, P. A., 127[39], *203*
Ahmadjian, M., 619[26,29], *622*
Ahmed, E. M., 654[73], *665*
Akamatsu, Y., 849[254], *872*
Akutsu, H., 848[224], 849[224, 254], 854[228], 855[228], *871, 872*
Alben, J. O., 724[17,18], 730 [17], *862*
Alberding, G. E., 654[72], *665*
Allen, G., 740[41], *863*
Allinger, N. L., 857[233], *871*
Allkins, J. R., 764[90], *865*
Almenningen, A., 249[63], *277*
Alvino, W. M., 920[83], *931*
Amat, G., 113[16], *202*
Ambrose, E. J., 163[74], *205*
Amy, J. W., 740[36], 791[36], *863*
Andermann, G., 395, *437*

Anderson, A., 54[25], 56[25], *69*, 736[31], 741[31], *862*
Anderson, D. G., 914[69], *930*
Anderson, D. H., 380, 430, 431 [125], *436, 439*
Angell, C., 244[41], *276*
Angelotti, N. C., 659[100], 660 [100], *666*
Angyal, S. J., 857[233], *871*
Appel, W. D., 703[2], *713*
Arnold, R. G., 649 650[56], *664*
Arnott, S., 822[193], *869*
Asato, G., 244[41], *276*
Asprey, L. B., 220[7], *274*
Attaway, J. A., 654[72], *665*
Aubke, F., 239[35], *275*
Avruch, J., 860[245], *872*

B

Bach, T. E., 364[17], 399[17], 400[17], 401[17], *435*
Bader, H., 625, *662*
Baier, R. E., 619, *622,* 997[57, 58,59], *1000*
Bailey, G. F., 453[15], *558,* 648, *664*
Baine, P., 284, 286, *344*
Baker, B. B., 603, *622*
Baldwin, S. F., 461[24], *559,* 621, *622*
Barbetta, A., 311[25], 312[25, 25e], 313[25e], 314[25d], 315[25d], 316[25d], 318 [25,25c,25e], 320[25c], *345*

Low, M. J. D., 456[20], 460[20], 558
Lozier, R. H., 764[91], 765[91], 865
Lundeen, J. W., 740[36], 791 [36], 863
Lundin, R. E., 654[61], 664
Luongo, J. P., 874[5], 889[19], 922[99], 927, 928, 931
Lutz, M., 764[93,94], 765[93], 865
Lyford, J., IV, 280[1], 281[1, 2], 282[1,2], 283[1,2], 284[1b,1d,2], 289[1,2], 290[2], 291[1,2], 292[1, 2], 293[1,2,15], 296[1,2], 300[1,2], 302[2], 303[1, 2], 305, 306[2], 308[2], 311[25,26], 312[25], 313 25b], 314[25b], 318[25], 344, 345
Lygin, V., 994[45], 1000
Lynch, P. F., 619[26,29], 622
Lytle, F. E., 432[134], 433[134, 135,138], 439, 440

M

MacGregor, D. R., 657, 666
Macintyre, W. M., 132[50], 203
MacKillop, D. A., 893[32], 929
Macleod, W. O., Jr., 654[71], 665
Maddy, A. H., 859[238], 871
Maeda, S., 392[60], 394[72,73 74], 395, 437
Magee, C., 243[39], 276
Maggio, M. S., 414[91], 438
Maher, V. M., 822[194], 825 [194], 869
Malcolm, B. R., 859[238], 871
Malm, J. G., 112[14], 126[38], 128[43], 202, 203
Malmstadt, H. V., 362[7], 434
Mammi, M., 908[53], 911[53], 930
Mank, G. A., 655[79], 665
Mann, D. E., 307[22], 345
Manning, D. J., 626, 662
Mansy, S., 790[163,164], 868
Mantell, C. L., 949[15], 950 [15], 998

Mark, H., 456[20], 460[20], 558
Markhart, A. H., 989[35], 999
Marks, H. B., Jr., 619[28], 622
Marquardt, D. W., 415, 438
Masakazu, H., 631[27], 663
Maslova, R. N., 827[198], 830 [201], 869, 870
Mason, A. A., 127[39], 203
Masri, F. N., 125[36], 203
Mathews, J. S., 657[96], 666
Mathieu, J. P., 118[24], 202
Matsuura, H., 764[259], 872
Mattson, J. S., 369, 435, 619, 622
Maxey, B. W., 281[3], 284[3], 286[10], 289[3], 291[3], 296[3], 310[3], 344
Mayer, A., 764[98], 865
Mayhem, I. R., 709[21], 714
Mazzacurati, V., 637[36], 663, 737[32], 862
McBride, A. C., III, 369, 435
McCall, E. R., 706[8], 710[8, 28,33,34], 711[35,37], 713, 714, 715, 990[38,39], 999
McCaulley, D. F., 659[101], 666
McClellan, A. L., 163[74], 205
McCormick, J. J., 822[194], 825 [194], 869
McCrea, D. H., 611[18], 622
McCubbin, T. K., Jr., 113[17], 202
McCullough, J. P., 468[30], 559
McDevitt, N. T., 213[4], 274
McDonald, R. S., 996, 1000
McDonald-Ordzie, P., 764[87], 808[87], 821[87], 833[87], 836[87], 865
McDowell, R. S., 220[7], 221 [11], 222, 252[13,67], 274, 277
McDugle, W. G., Jr., 239[34], 275
McFadden, W. H., 654[61,70], 664, 665
McFarlane, W., 246[51], 276
McFee, R. H., 584[8], 593[8], 621
McGinness, J. D., 954, 998
McGlothlia, R. E., 268[76], 277

Part A, pages 1-346; Part B, pages 347-716; Part C, pages 717-1000

CKKD

Date Due

E

PRINTED IN U.S.A.   CAT. NO. 24 161   BRO DART